the art of science

Activities and creative ideas for the teaching of science to infants and lower juniors

Barbara Hume and Christine Galton

Illustrations by Kathie Barrs

First Published in 1989 by
BELAIR PUBLICATIONS LIMITED
P.O. Box 12, Twickenham, England, TW1 2QL

Series Editor Robyn Gordon
Designed by Richard Souper
Photography by Kelvin Freeman
Typesetting by Woodhead Designs Service Limited
Printed and Bound by Heanor Gate Printing Limited
ISBN 0 947882 11 1

Acknowledgements

The authors and publishers would like to thank the children and staff of East Sheen Primary School, Richmond-upon-Thames and Orleans Infants School, Richmond-upon-Thames, for their co-operation in the making of this book, and Katie Kitching for contributing the cover artwork.

They would also like to give special thanks to Shirley Hampson in her role as Science consultant.

Contents

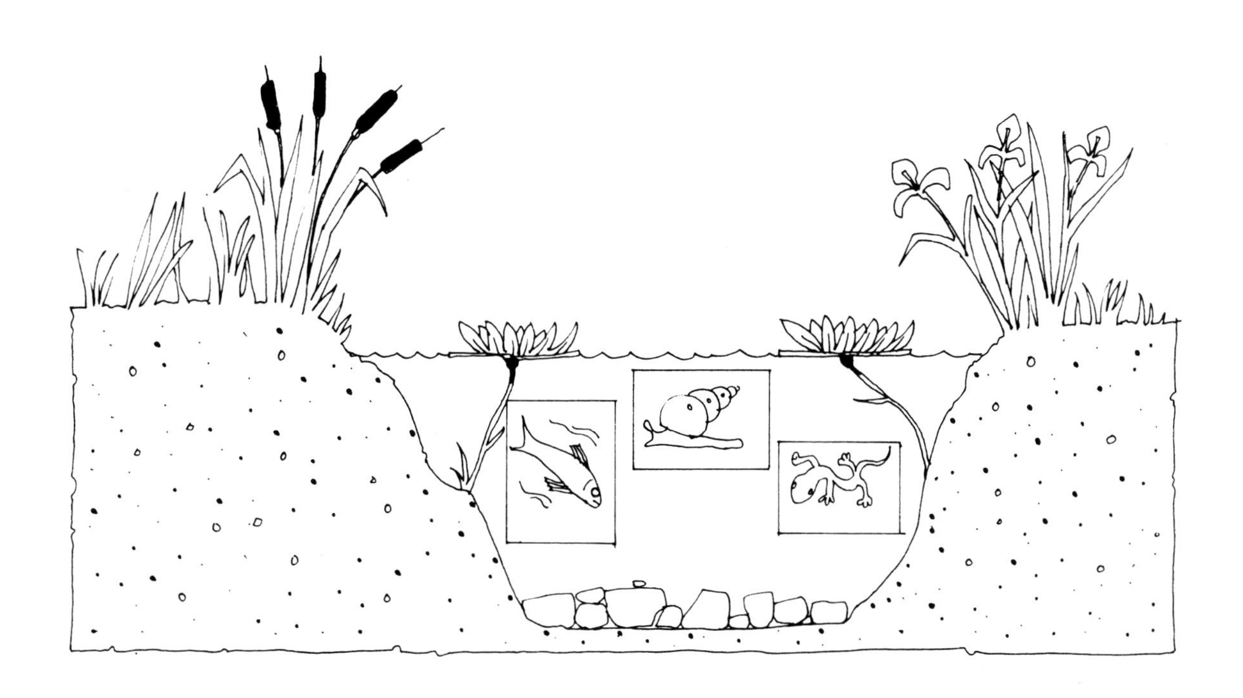

Introduction

In *The Art of Science* we aim to provide a resource book of ideas for both science and related creative activities in the infant and lower junior school.

The ideas are presented within twenty-eight topics. However, each topic need not be independent of the others. For example, the study of Toys might include work from several areas such as Magnets, Floating and Sinking, Wooden Things or Batteries and Bulbs. The storybook approach might lead to investigations from many different areas of science. Some starting points, together with how they may be developed, have been suggested in the final chapter, Science from Stories.

Each topic includes starting points, areas for discussion, questions that may be asked to stimulate investigations, and a list of relevant vocabulary. There are ideas for collections and how they might be displayed to stimulate interest and curiosity; for stories that reinforce or introduce concepts, and for visits and opportunities for direct observational work.

We believe that in carrying out their science work, children should be encouraged to discuss their ideas, to make observations and to reach conclusions. Through their investigational work the children will begin to appreciate the need for 'fair tests', to try out their own ideas, to predict what might happen, to determine the materials they will need, and communicate their findings. We hope that the children will develop a respect for all living things and an awareness of environmental issues.

⚠ Care has been taken to highlight areas that need special attention on the part of the teacher to ensure the safety of the children carrying out investigational work. In fact, it is fully intended that science should have a positive role to play in health and safety education.

We should like to stress that activities involving tasting should be avoided unless they are carried out within the context of cooking, when food should be stored and prepared under hygienic conditions.

We would like to emphasise the benefits of reinforcing science with art:
– Many of the suggested creative activities, for example wax resist techniques, colour mixing, and dyeing, involve a scientific process.

– Creative activities enable the children to record their science work through pictorial representation and observational drawings.

– In addition, through creative and craft work, the children's skills of designing and making are developed.

It is hoped that the topic approach will enable children to apply their findings and relate their scientific discoveries to real-life situations.

Finally, we hope that the ideas presented in this book will encourage both teachers and children to develop as interested, creative investigators of science.

Barbara Hume and Christine Galton
1989

Colour

Starting Points/ Discussion: Talk about colour associations, e.g.

RED: robin, fire, blush, cherry, poppy, ruby, 'red as a beetroot'.
YELLOW: sunshine, sand, honey, straw, lemon, canary-yellow, 'yellow as butter'.
GREEN: grass, jade, emerald, apple, olive green, pea green, bottle green.
BLUE: sky blue, egg-shell blue, peacock blue, kingfisher.
WHITE: frost, snow, alabaster, milk, flour, swan, pearl.
BLACK: coal, charcoal, soot, ebony, jet, ink, blackberry.
GREY: pewter, slate, ashes, dove-grey, mousy, steel-grey.
BROWN: bronze, coffee, chocolate, caramel, toffee, burnt almond, walnut, tan.
ORANGE: brass, tangerine, marmalade, ginger.
PINK: salmon, shell-pink, raspberry.
PURPLE: lavender, violet, mauve, plum.

Look at the names on paint charts. Make up some new ones.

Talk about favourite colours: colours associated with festivals and seasons (e.g. autumnal browns, red and green at Christmas, yellow and white at Easter, red for Chinese New Year, colour associated with weddings around the world); talk about colours and markings in nature, e.g. bright colours often mean DANGER; camouflage; look at eye colours, hair and skin tones within the class; consider colours used for signals, signs and coding, e.g. traffic lights and wiring a plug; look at colours of children's clothes and if possible sort them into sets.

Vocabulary: Shades, tones, tints, hues, light, dark, pale, bright, vivid, vibrant, dazzling, fluorescent, faded, dull, 'washed-out'.

Collections: Make colour tables – items of the same colour gathered together. Paint charts; wallpaper, carpet, material sample books; packets of felt-tips, wax crayons, etc. that show a range of colours; wools and threads; Cellophane, tissue paper, coloured acetates; balloons, tinted plastic bottles, sun glasses, visors, etc. for looking through; boxes or packets of items used to tint, colour and dye, e.g. hair tints and colourants, dyes, food dyes, stains, make-up, shoe polishes and dyes; posters and postcards of paintings showing use of colours.

Colour

- **Make a colour chart or word bank:** colour names can be detached and re-matched, e.g. lollipop shapes in a sweet jar.

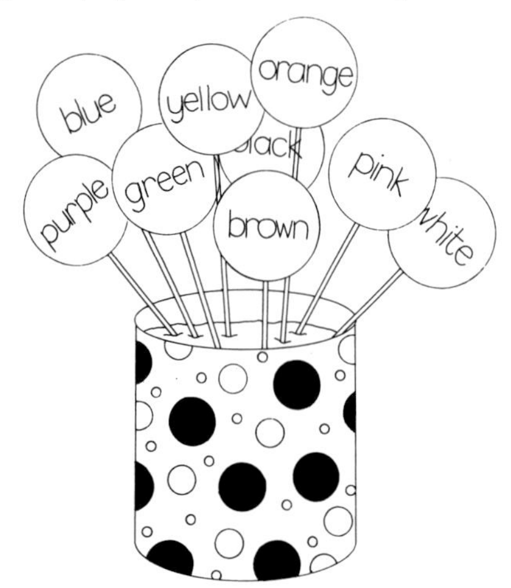

Science Activities:

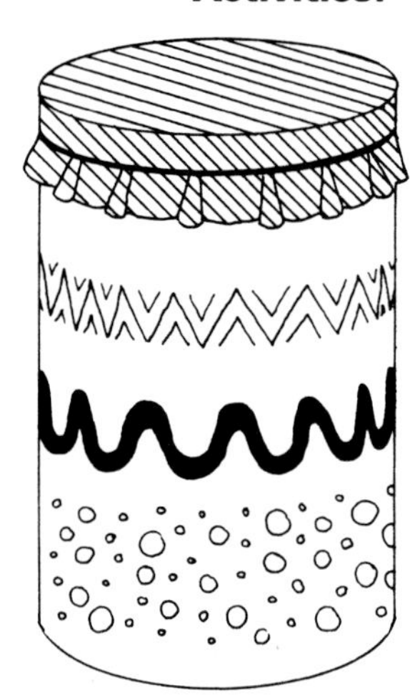

- Look for rainbow colours produced by prisms, bubbles and in puddles.
- **Make viewers** ('lookers') out of cardboard tubes and Cellophanes. Paint tubes, fix Cellophane with a rubber band first to gather it; secure with masking tape or coloured adhesive tape. How can you make darker coloured viewers (use more than one layer of Cellophane)? Can you look through more than one viewer at a time? Shine a torch through your viewer onto a piece of white paper. Can you find other things to look through? Can you change colours? (Try a bottle, a balloon, sun glasses.) Draw a picture, write your name with a coloured felt-tipped pen. Observe through a viewer.
- Put different colours of Cellophane on torches. Can you mix the colours on a screen? Try red, green and blue.
- Foods:
  - Which foods change colour when cooked?
  - Make rainbow biscuits and cakes. Investigate which colours are most acceptable in food.
  - How does dilution change the colour of liquids?
  - Look at the colours used in the packaging and advertising of different foods.
- Make coloured spinners. Try different colour combinations.
- Investigate the colours of living things, e.g. fish, birds, minibeasts, flowers, leaves and berries.
- Chromatography: some inks are a mixture of colours. Mark a piece of paper towel or filter paper with black ink or a black felt-tipped pen. Drop water onto the ink and watch the colours separate. When the paper is dry add details and make the shapes into pictures.
- Dyeing: using both commercially produced dyes and vegetable dyes, e.g. red cabbage, onion skins, investigate fixing agents – e.g. salt. Try hot and cold water. Try over-dyeing (one colour on top of another). Try different materials with the same colour.
- Fading: investigate fading using coloured papers, e.g. crêpe, tissue, art paper. Put papers in different positions – on a window ledge, in a cupboard – to determine how fading occurs. Test materials, papers and perhaps sweets for colour fastness. Try hot and cold water.
- Investigate contrasting colours: which are the most effective colour combinations?

Colour

Creative Activities:

- Mix paints. Primary colours of paint are red, blue and yellow. Use tempera or powder paints and mix on a palette or on the paper (see photograph on p.5).
- Mix tints and shades of one colour by adding white paint or black paint systematically. Make a tint or shade chart. Paint a picture or pattern with the tints and shades.
- Design a colour scheme for a model 'room in a box', or for a paper doll's outfit.
- One-colour collage. Collect materials, wools, papers, beads and buttons of different shades of one colour. Arrange materials to make a pattern or picture (spirals, stripes, concentric shapes).
- Make tissue paper or Cellophane window frames. Look at the colour of the overlaps.
- Make a black and white display.
- Roller printing. Paint the roller with a brush. Put several colours in bands on the roller.
- Weave ribbons, wools and material strips of different shades of one colour.
- Design a safety warning poster using the most effective colour combinations for attracting attention.
- Illustrate a 'colour' story such as *Brown Bear, Brown Bear, What Do You See*? by Eric Carle, Picture Lions or *The Mixed-up Chameleon* by Eric Carle, Picture Puffins. Make up and illustrate your own colour story.
- Organise a 'rainbow stall' for your summer fête. Sort items for sale according to colour and place on tables covered with crêpe in the colours of the rainbow.

Toys

Starting Points/ Discussion: The topic of Toys provides ample opportunities for craft, design and technology and for the study of materials, energy and moving things. The topic can be investigated at any time of the year, but may be introduced by Christmas or a birthday celebration. Talk about the children's favourite toys. Carry out a toy survey in the class to find out the most popular, e.g. how many children have a dolls' house, train set, model cars or a teddybear? A visit to a toy museum may initiate an investigation into old toys. Which toys are still popular today? Compare old toys with their modern equivalents. How are they different? Discuss how toys could be sorted out, e.g. construction toys, board games, 'cuddly toys', wheeled vehicles, toys that lift, roll, rock, jump, swing or spin. Investigate toys from different countries.

Vocabulary: Wheels, axle, pulley, cog, gear, drive, chassis, level, spring, balance, jointed, puppet, top, construction toys, yo-yo, pogo stick, rocking horse, skateboard, clockwork, jack-in-the-box, hobby horse.

Collections: Most popular toys, old toys, toys from different countries, dolls, puppets, soft toys, construction toys, games, jigsaw puzzles, wheeled toys, magnetic toys, water toys, battery powered toys, models, a baby's toys, balls, etc.; pictures from catalogues.

Science Activities:

- Investigate the material from which toys are made, e.g. wood, plastic, metal, fabrics, rubber, paper. Sort out toys according to the materials used. Consider the safety aspects of toy manufacture, e.g. there should be no sharp edges, dangerous points or splinters; non-toxic paints and finishes should be used and there should be no detachable parts that may be swallowed. Ask the children to bring in toys that they had when they were babies. Look at the materials used for babies' toys. Why are they especially suitable? Look at the recommended age ranges for different toys. Investigate which materials are the most hard-wearing. What signs of age do toys show? Metals may rust, wooden and plastic toys may be scratched, fabrics may show signs of wear.

- **Investigate wheeled toys.** Ask children to bring wheeled toys to school, e.g. bicycles, roller skates, skateboards, scooters, prams, toy cars and lorries. How do they move? How are they powered, e.g. battery, clockwork, human propulsion? Make wheeled models from construction toys, e.g. Lego, Meccano. How does your model vehicle travel on different surfaces? Try rough and smooth surfaces, wet and dry, flat and sloping ground. Which vehicle travels furthest? Devise a fair way of measuring, e.g. by letting go of them at the top of a ramp. Alter the shape and weight of your vehicle and see how far it travels. Pull your vehicle up a slope. How can you measure how much energy you need (attach a rubber band or elastic, or use a spring balance)? Change the angle of the slope and observe what happens. Make wheeled shoe box vehicles. Consider the best materials for the wheels. What problems do you meet with cardboard? What happens if the wheels are not perfectly round? Try plastic lids. What happens if the axle doesn't go through the centre? How can you find the centre of the lid? Look for pulley wheels, gear wheels (bicycle, clocks, egg beaters). Investigate drive wheels.
- Investigate balls. Collect as many different kinds of ball as possible, e.g. football, rugby, golf, tennis, billiard, polystyrene, cricket, bowling, ping-pong, sponge, rubber, inflatable beach ball, marbles, plastic 'airflow' ball, etc. Compare them, e.g. size, weight, shape, texture. Investigate how the balls roll. Which balls are the best for bouncing? Look at angles of rebound. Which balls are used with a bat? Match bats (cue, club) to balls. Look at the materials used. Make a skittle game or a marbles game.

Toys

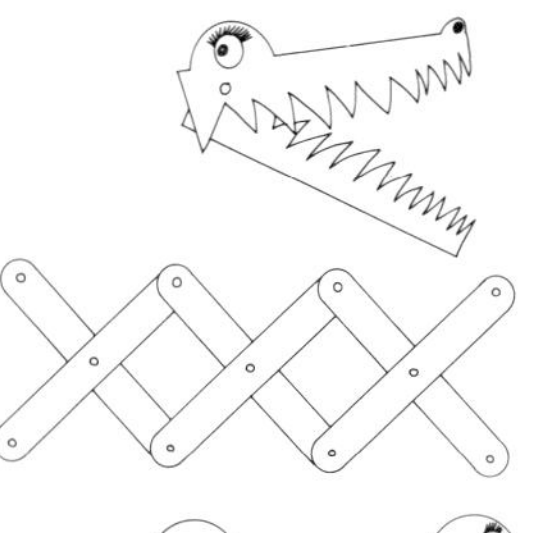

- **Investigate toys with 'joints',** e.g. puppets and dolls. Collect jointed toys and look at the way in which they can move. Make jointed shapes with Meccano and geostrips – experiment with different ways of making them move, e.g. make the hands of a paper clock go round, make tongs and scissors from geostrips, make signals, a jointed snake or a caterpillar.
- Water toys: investigate toys that float, make a submarine and boats. Investigate water squirters, syringes, pumps, water-wheels.
- Balancing toys: make spinners, tops, mobiles, high wire acrobats.
- Make models of things that lift, e.g. a crane with a winding mechanism, a pulley to hoist a flag.
- Make magnetic toys and games (see METALS AND MAGNETS).
- Light up models and toys with batteries and bulbs.
- Make gliders, shuttlecocks, windmills, toy parachutes, kites and paper spinners.

Creative Activities:

- Close observational drawings of old and new toys.
- Design and make a jigsaw puzzle.
- Make shoe box wheeled vehicles. Use either the box or the lid.
- Make shadow puppets.
- **Make jointed 'string-pull' puppets** (Father Christmas, teddybear, penguin).
- Make string puppets (see photograph above).
- Make a painted toyshop frieze.
- Make a 'stored energy' toy, e.g. a catapult.

It's party time!

Starting Points/ Discussion: The starting point can be a birthday party, a festival or celebration such as Christmas. Talk about plans that need to be made. What do you need for a party? Food, decorations, invitations, presents and cards, tableware, entertainment, party bags, etc. Read *The Trouble With Jack* by Shirley Hughes, Picture Lion.

Vocabulary: Balloons, rubber, blow up, inflate, deflate, pump, pop, burst, static electricity, puncture, jet, stretch; liquid, solid, jelly, freeze, melt, dilute, dissolve, set, fizzy, carbonated.

Collections: Party tableware (paper and plastic plates, cups, spoons), streamers, blowers, balloons, balloon pump, party hat, party bags, invitations, birthday cards, different types of wrapping paper, ribbons, string, adhesive tape, cake candles and holders, wrappings from favourite party foods, jelly moulds, cake tins, containers for preserving food, cling film, foil, plastic and paper bags.

Science Activities:

Balloons

- Look at the colour, shape, size and stretchiness of a balloon before inflation. Can you predict the shape of the balloon after inflation? How can we inflate a balloon? (Use balloon pumps). Look at the skin of an inflated balloon. Is it shiny or dull? Has it changed in colour?
- Consider what will happen if you let it go. Let go and look at the path a balloon takes as it deflates. Listen to the noise it makes. Look at the deflated balloon. Is it exactly the same size as it was before inflation? Can you devise a way of measuring the difference in size (e.g. by drawing around it before and after)?
- Consider how you can tie a balloon to stop the air escaping. Devise a fair test to determine the best method. How can you make sure that the balloons are inflated to the same extent?
- Investigate static electricity produced by rubbing the skin of the balloon. Can you make the balloon 'stick' to the wall? Can you make things 'stick' to the balloon? Investigate surfaces for rubbing.
- Investigate an inflated balloon over a period of time. What changes can you see?
- How many different noises can you make with a balloon? Don't forget BANG!
- Fill a balloon with water. What happens when you pop it (in a sink or in a water tray)?

Food

- Make jelly. What is needed to make jelly cubes turn to liquid? What happens as the jelly mixture cools down? Which is the best place for setting jelly quickly? Try the fridge, on top of radiators, in a warm room, etc. Can you devise a fair test?
- Make biscuits: observe differences between cooked and uncooked mixture. What has the heat of the oven done to the biscuit mixture?
- Make sweets, toffees. This provides the opportunity for looking at different types of sugar, for sieving, for observing changes in consistency, drying-out and setting.
- Make ice lollies by freezing a mixture of water and flavouring. How can you fix the lolly-stick? **Make real lemonade.**

- Preserving foods (party food has to be prepared in advance): how can you keep your food fresh until your guests are ready to eat? Investigate how to avoid soggy crisps, hard cakes, melted ice-cream, flat lemonade and curled up sandwiches.
- Look at wrappings, e.g. cling film, greaseproof paper, foil and containers.

It's party time!

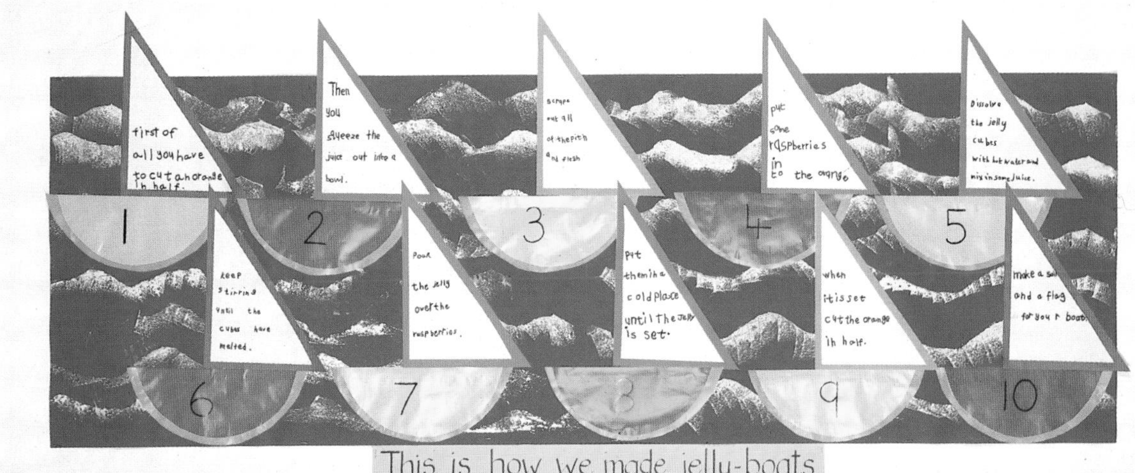

Illustrate a recipe.

Parcels and presents

- Investigate different ways of fixing wrappings, e.g. adhesive tape, string, ribbon, glue, staples. Find out about different ways of knotting string.
- Investigate ways of wrapping a breakable present, e.g. by using 'bubble' packaging, polystyrene pieces, tissue paper, cardboard, etc.

Entertainment

- Make music for party games.
- Shadow puppet show (see SHADOWS).
- 'Magic' tricks, e.g. involving magnetism.

Creative Activities:

- Invent a birthday card or party invitation with moving parts, e.g. make a jack-in-the-box pop out. Use flaps, levers, hinges and springs.
- Make printed patterned wrapping papers. Suit the techniques used to the quality of the paper: vegetable prints, marbling, stick printing, dip paper towel, wax resist, etc.
- Investigate the best paper for weaving, curling, tying bows, making streamers, etc.
- **Design a party bag and decorate it** (consider how strong it has to be and whether it is to hold food).
- Make clay or dough pretend party food for the playhouse. Paint and coat with PVA glue.
- Balloons: – make papier-mâché balloon creatures.
  – **decorate balloons**, e.g. make faces, give them hair, hats, feathers, etc. Attach shirring elastic.
- Make and decorate a party hat, a table mat, serviettes and table cloth.

Shining Things

Starting Points/ Discussion: Starting points might be a Christmas tree bauble, a mirror, some shiny new shoes, a collection of conkers or leaves, or a shiny-leaved plant. Read 'Shining things' by Elizabeth Gould from *The Book of a Thousand Poems,* Evans Bros. Shine the light from a torch onto reflective surfaces, e.g. a tin lid or a mirror, and 'bounce' the reflections around the room. ⚠ Look at reflected light from a prism. **Not thin glass as this shatters dangerously.** Look for the colours of the rainbow in these reflections.

Vocabulary: Bright, glittery, glinting, sparkling, dazzling, glistening, glossy, sheen, reflection, shimmering, brilliance, luminosity, radiance, lustre, glare, burnished, glassy, gleaming, twinkling, glazed, varnished, polished, dull, dim, gloomy, sombre, sooty, murky, misted, cloudy, matt, lacklustre, filmy, muddy, rusty, grimy, smeared, mildewed, dirty, dusty.

Collections: Ask children to bring in (with the help of their parents) 'something you can see your face in'. Collect and display: tinsel, decorations, foils, bottles, beads, mirrors (plastic safety mirrors), sequins, shiny fabrics, plastics, metals, patent leather, prisms, shiny conkers, leaves, stones, glazed pottery, plastic macs, Wellington boots, etc. Collect *packaging* of polishes, e.g. shoe polish, furniture polish; black and white paper.

Put some stones in water on the display table; put some bubble-mixture, and a balloon with a balloon-pump on the table too. Safety clothing with reflective bands or strips.

Science Activities:

- Sort collected objects according to material: e.g. wood, metal, plastic, glass, fabric.
- Investigate reflections: mirror-writing, symmetry, images in convex and concave mirrors. **Investigate two hinged mirrors:** how many images can you see? **Make a kaleidoscope with three mirrors.** Make a periscope. Invent a device for looking round corners.

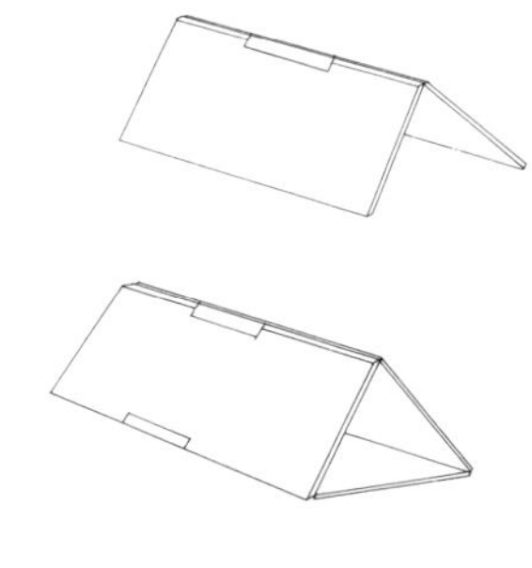

- Look for 'reflectors' (for both heat and light). When are they used?
- Experiment with the beam of a torch. What do you notice about the area around the bulb? Play with the reflections, bouncing them around the room – how can you alter the shape of the reflections? What kind of surfaces make the best reflectors? Does the colour matter? Do transparent things reflect?
- Can you make a shiny thing dull? Try smearing a mirror with grease and dirt, paint, talcum powder or flour. Test by shining a torch beam on the mirror and looking at the reflection. Can you make a dull thing shiny? (See Creative Activities.) Polish shoes, wood, silver, windows, mirrors.
- Test shiny surfaces for water repellent qualities. Are all shiny things waterproof? (Try satins, silks, as well as plastics and metals.) Why do we glaze pottery, or paint woods and metals with gloss paint?
- Look for shiny things in the natural world, e.g. leaves, bark, berries, insects, pebbles, minerals, crystals.
- Test shiny things for 'slipperiness'. Make a good slope for a slide. Compare rough and smooth surfaces, shiny and dull surfaces. Consider the angle of the slope, the shape of the object. Does it roll or slide? **Pull a block along a surface using the weight method over the edge of a table.** Attach a tube to the edge of the table to allow the string to run smoothly.

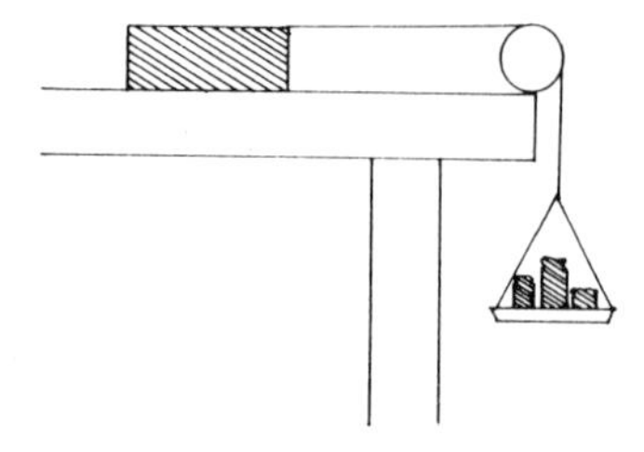

Shining Things

Creative Activities:

- Use contrasting shiny and dull paper/fabric strips to weave a mat.
- Make collages using shiny papers and materials, Cellophane, foils, sequins, buttons, tinsel, lametta. Frame with strips of foil stapled at the corners (use mini-stapler to fix) — or glue.
- Glazed tissue pictures. Stick tissue shapes to backing paper using PVA glue spread liberally. Spread a layer of glue over the top for a glazed effect. (See photograph above.)
- **Beaded sewing.** Use Indian mirror-work as a stimulus. Mixture of sewing and collage using sequins, gold and silver thread, shiny fabrics. Sew on to dull fabric for contrast.
- Take rubbings using gold, silver and bronze crayons. Rub over the same object several times with three different colours to make a repeated pattern. Arrange in checked pattern or as a triptych. (Try with card rubbings too.)
- Paint and varnish pebbles.
- Glaze clay models or apply PVA glue to painted clay or dough models.
- Make shiny foods in cooking, e.g. jelly, crystallised fruit, glazed pastry and breads, toffee apples.
- Shiny mobiles. Make star or moon shapes of foil and hang from tinsel covered hoops. (One side of stars could have vocabulary written on.)
- Make close observational drawings of reflections in shiny objects, e.g. in mirrors, spoons, kettles.
- Symmetry: – make symmetrical patterns on squared paper
  – ink-devil prints
  – symmetrical cut-out patterns by folding and cutting.

Clothes

Starting Points/ Discussion: Look at the clothes you are wearing. How many layers are there? What are the fabrics? How are the clothes fastened? Are all the jumpers the same — are some hand-knitted? How can you tell? Which clothes are stretchy, strong, thick, warm? What would you do if you were too hot — too cold? Look at your coats — which are waterproof, padded, etc.? Why do we wear gloves, mittens, scarves or hats? Look at P.E. kits. Why do you need to change for P.E.? Look at clothes for different purposes — rainwear, beachwear, winter clothes, summer clothes, play clothes, 'best' clothes, sports clothes, safety clothes (reflective, bright colours as well as crash helmets, knee pads, etc.). Read *You'll soon grow into them Titch,* by Pat Hutchins, Picture Puffin. Look at adults' clothes and compare with children's clothes or baby clothes. Look at garment labels. What information do they give? How do we get clothes clean? Look at natural and man-made fabrics. Look at clothes from different cultures and climates.

Vocabulary: Waterproof, windproof, washable, tough, padded, lined, fur, leather, cotton, polyester, nylon, wool, knitted, woven, warp, weft, texture, stretchy, fabric, zip, buckle, button, buttonhole, Velcro, hook and eye, popper, press stud, laces, toggle, elastic.

Collections: Sets of children's clothes; clothes from different cultures and climates; sports clothing; protective clothes, e.g. skateboarding or motor cycling; clothes for special occasions: bridesmaids' dresses, Christening gowns; clothes made from different fabrics; clothes which are old — how have fashions changed since your parents were little? Collect hats and display on balloons; catalogues for clothes; natural raw fibres: leather, suede, fur, etc; knitting wool and needles (big wooden ones); a simple paper pattern to show how many pieces a shirt has; sewing machine (take needle out); small weaving looms.

Science Activities:

- Which fabrics are waterpoof? Start by collecting some different fabrics — preferably old clothes or coats. Decide on four fabrics to test. How can you test the fabrics fairly? The fabrics can be stretched over jam jars and secured with elastic bands. Spray water onto the top of each jar (a plant sprayer is a good source of 'rain'). Prepare a chart to record the results. Results will depend on the fabrics chosen: some will soak up the water, some will cause droplets to form and some will let the water pass straight through.
- Which liquid makes the clothes the cleanest? Experiment with different liquids used for washing — liquid soap, liquid detergent, washing-up liquid. Investigate this further by trying hot or cold water with the liquid washing agent. Wear rubber gloves.
- Investigate the drying of clothes. Hang up pieces of wet cloth. Which dries the quickest?
- How hard-wearing are different fabrics? From a collection of old clothes or pieces of fabric select four to test for durability. Which fabrics do you think are hard-wearing and why? Discuss different ways of testing, e.g. sandpaper blocks, scratching with stones.
- **Design playclothes using the information gained from the above investigation.**
- Which fabrics make the warmest clothes? Devise a fair test.

Clothes

- Test fibres/threads for strength and elasticity.
- Test for strength of different fastenings — Velcro, poppers, hooks and eyes, buttons.
- Use a magnifier to look closely at weave and textures of fabrics.
- Investigate the absorbency of different fabrics (e.g. cotton, nylon, wool, polyester).

Creative Activities:

- Dyeing — use natural dyes, e.g. red cabbage, blackcurrant, onions, etc. Tie and dye — old T-shirts, **use commercial dyes with care.**
- Collage — use fabric pieces to make pictures. Encourage the use of different textures, prints, weaves. etc.
- Make peg dolls or wooden spoon dolls.
- Rag printing — use different fabrics to give different effects: net, nylon, fur fabric, etc.
- Print on to fabric using vegetables or blocks. Use fabric dyes, or paint if it is to be purely decorative.
- Observe patterns in printed fabrics as stimulus for own pattern-making.
- Paper/card weaving.
- Collage using pictures of clothes from catalogues. **Make a 'top and tails' book.**

Candles

Starting Points/ Discussion: A good starting point for investigations into candles could be a festival, e.g. Hanukah, Divali, a Christingle service, a birthday or visit to a church. What do candles feel like? Can you dig your finger nail into the surface or scratch it with a pencil? When do we use candles (festivals, celebrations, power cuts)? Why were candles important in the past? What are candles made from? What other uses do we have for wax? Discuss waxed fabrics, papers, sealing wax, wax crayons, waxed thread (used by leather workers), modelling wax, waxing skis and surfboards. What do you think the wick is made from? What does it do (draws up inflammable liquid to the flame)? What else behaves like wax (butter, metals, soap, jelly, chocolate, glass)?

Vocabulary: Tallow, taper, spill, nightlight, wick, wax, glow, soot, melt, liquid, solid, flame, snuff, extinguish, candlelight, candelabra, candlestick, wax chandler, wax doll, wax works, wax impression, resist, mould, 'soft as wax' 'burn the candle at both ends', (display words on candle-shaped cards).

Collections: Collect as many different candles as you can. Sort them for display according to shape, size, colours, scented/unscented, decorated/undecorated; candlesticks and holders; sealing wax, polish, beeswax, wax crayons, wax doll.

Science Activities:

⚠
- Observe a burning candle. What do we need to light a candle? What shall we put the candle in (foil case filled with sand stood on a biscuit tin lid or metal tray)? **Discuss how candles can be dangerous and what kind of things burn** (hair, skin, clothing). Which part of the candle do we light? Light different coloured and shaped candles (e.g. birthday cake candles, nightlights, household candles). Look at the flames. Are they all the same size, shape, colour? Are the flames still or moving.
- Collect soot from the candle flame on a cold metal teaspoon.

⚠
- **Look at the shape of the flame when you blow gently.** Can you see anything above the flame? (Carefully feel the warm air well above the candle.) Can you smell anything? Can you see any smoke? What is happening to the wax? Where is it melting first? Can you see a dip in the candle around the wick? What happens to the wax that drips down the sides? Look at candlelight in the darkness if possible. Collect words to describe the quality of the light. How can we extinguish the flame. Consider candle snuffers. What has happened to the wick? Is it standing upright? Has it changed colour? Does it carry on smoking.
- Weigh the candle before and after burning, using standard or non-standard units of weight.
- Burn a nightlight under a wide-necked jar. How long does it take for the flame to go out? Change the size of the jar and time it again, to compare. (This demonstrates that something in the air is being used up as the candle burns.)
- **Design a safe candle-holder.** Investigate stability, suitable materials, shape, how the candle is held. Make Divali lamps with clay, coloured dough, plasticine or plaster of Paris.
- Investigate wick action using jars containing a small amount of water. Test strips of fabric, paper towel, blotting paper, string (unwaxed), cottons, wool, shoe laces. Colour the water to make readings easier.
- **Make wax candles under close supervision** (by melting wax crayons) in egg-box sections or yoghurt cartons. Vary colour, diameter, height, thickness of wick.
- Test wax as a lubricant (waxing drawer runners, etc.).
- Make a candle clock.
- Extend investigations to include other substances that melt when heated, that can be moulded and that solidify on cooling, e.g. chocolate.

Candles

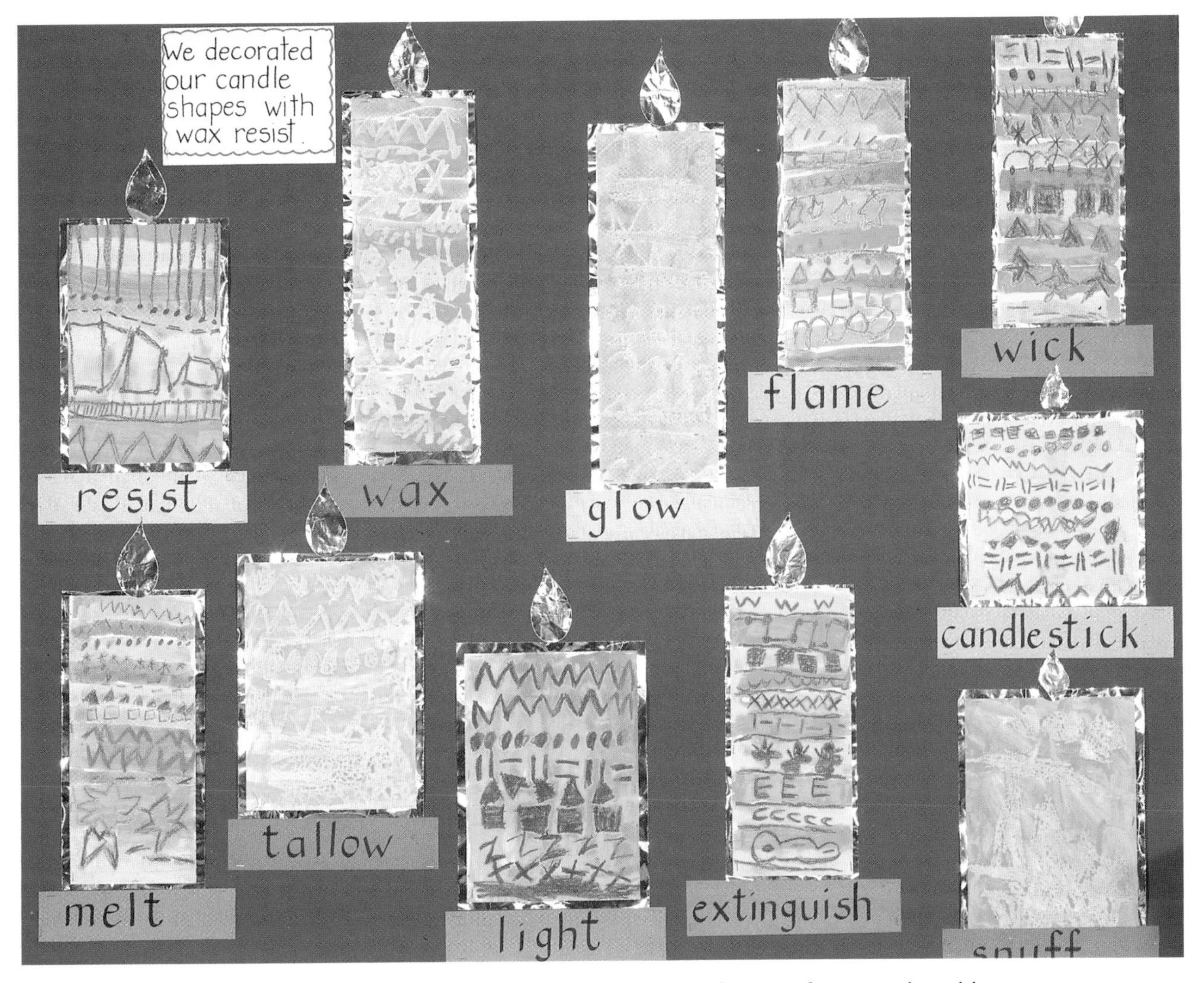

Creative Activities:

- Crayon patterns onto a candle-shaped (rectangular) piece of paper using either wax crayons or the sharpened end of a candle. Brush over your design with a coloured wash using inks or powder paints. (See photograph above.) Use your wax-resist candle shape as a basis for a candle collage picture.
- Make 'magic' pictures. Crayon heavily all over a piece of paper using different colours. Turn paper over and place on a white piece of cartridge paper. Draw a detailed picture using a sharpened pencil, ball-point pen or modelling tool, pressing firmly.
- Wax rubbings. Rub over textured surfaces, coils of string, cardboard shapes, bark, leaves, scissors, coins, rulers. Brush an ink or powder paint wash over the rubbings.
- Cut/scrape pictures into wax crayon. Cover a piece of paper with haphazard shapes and different colours of wax crayon. Cover this with a layer of black crayon. Cut through the black layer with a clay tool or sharpened pencil to reveal the colours underneath. This is a good technique for depicting bright lights in the dark.
- Crayon and ink cutting. Cover the paper with coloured inks; crayon over and then scrape through the crayon to reveal the ink colours.

Allow plenty of opportunities for experimenting with the above techniques. Let the children discover how wax, water paints and inks interact and the effects that can be achieved. Vary the colour of the washes according to the topic, e.g. blue/green for the sea, white for snow, black or dark grey for night scenes.

Candles

- Make close observational drawings or paintings of candle flames.
- Make 3D representations of candles with kitchen roll cardboard tubes decorated and painted. Add Cellophane/tissue paper flames.
- **Make candle mobiles**
  Hang on a hoop covered with crêpe paper. The flames turn around independently of the candle shapes.
- **Pasta collage of candle shapes**
  Make flames with foils and Cellophane. The candle shape can be decorated with bands of pasta shapes, red lentils, seeds etc. or with cut out wax-resist patterns.

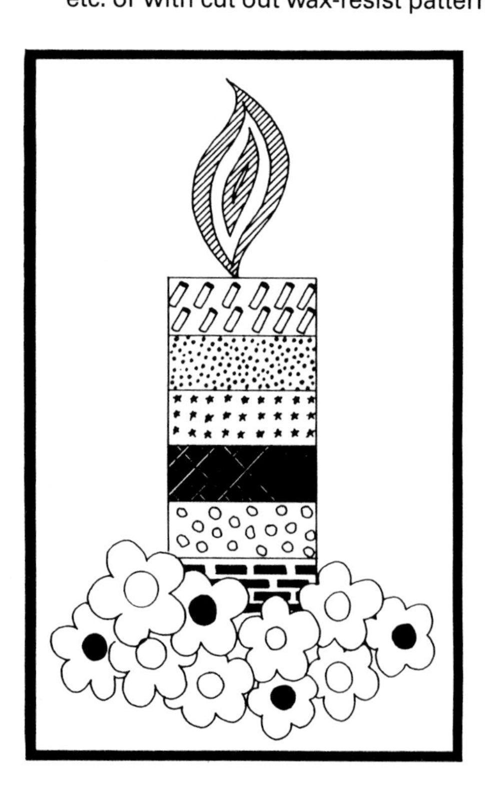

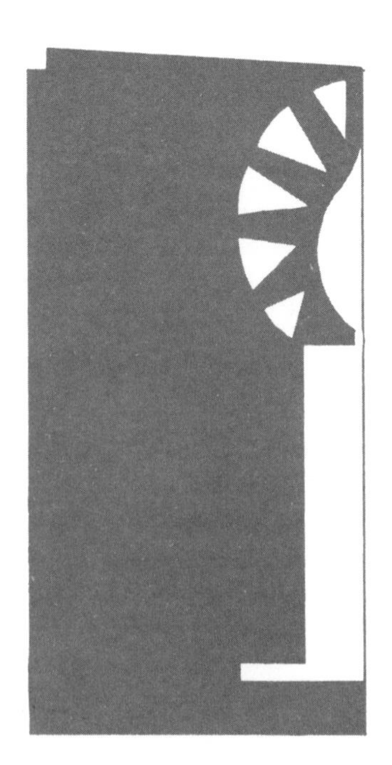

- **Make a candle tissue paper window frame.**
  Fold a piece of paper in half; cut out half a candle shape on the fold. Glue tissue shapes across the cut-out (strips and circles, over-lapped or torn shapes are all very effective).

Leaves

Starting Points/ Discussion: The starting point for the discussion on leaves will depend upon the time of year. In spring the topic could be incorporated into a study of growth; in summer the focus could be on leaf shapes and patterns, or minibeasts that eat leaves; in autumn falling leaves can be observed, while in the winter months the focus could be on the use of evergreens as decorations for winter festivals. The study of leaves throughout the year could also be an integral part of an 'adopt a tree' topic.

⚠ **Safety: Some leaves are poisonous. Sap can irritate the skin. Wash hands after handling leaves.** Some leaves are as sharp as needles! For the above reasons, we strongly suggest that a study of edible leaves (leafy vegetables and herbs) should be specifically undertaken within the context of 'cooking'.

Vocabulary: Stem, vein, subdivision, leaflet, shoots, scar, leaf-litter, leaf-miners, propagation, cutting, serration, skeleton, tendrils, buds, perennials, annuals, evergreen, deciduous, seedling, foliage, frond, blade, needle, prickle, thorn, succulent, cactus, rosettes, insectivorous, strap-like.

Collections: Leaves: stress to the children that leaves are *vital* to the well-being of plants and trees and that great care must be taken not to damage living things. Observe leaves on plants, collect leaves that have fallen — perhaps after a strong wind or storm. Take advantage of prunings, or selectively cut a twig or two for studying indoors. Bring in potted plants. There is now a huge variety of houseplants that show the vast range of leaf colours, shapes and arrangements. Beware of some that are poisonous. Display pressed leaf collections and leaf/ flower presses. **A collection of green things can be displayed.**

Leaves

Science Activities:

- Observe leaves on plants and trees. Look at their shapes. Are they strap-like, broad, flat or like needles? How do they feel? Compare the under-side with the top surface. How does the plant hold the leaves to gain the most light? Look at the pattern of the veins, at the distance between each leaf on a stem and how they are arranged around the stem. Can you see any leaf scars? How much shade do the leaves give? Observe leaves in the rain. Look at the tips of the leaves. What happens to the water on the surface? Watch leaves in the wind. Do some leaves move more than others? Look for tendrils.
- Sort collected leaves according to shape, size, colour, serrated/non-serrated edges, shiny/non-shiny, variegated or not, number of leaflets.
- Investigate the length and width of a common leaf, e.g. privet or dandelion. Order leaves according to length and make a chart. Introduce the idea of sampling. Choose plants growing in different places and make comparisons. Can you see any patterns in your results? Compare leaves grown in the classroom, e.g. fast-growing plants such as mustard, cress, sunflowers; compare leaves of seedlings with those of mature plants.
- **Grow leaves from the tops of turnips, parsnips, radishes, pineapples.**

- What do leaves need to grow and to stay healthy? **Does water reach the leaves of plants?** Try standing celery stalks in water (use red or blue food colouring). Cut the ends of the celery stalks (choose stalks with leaves). Leave for several hours (or overnight) and watch your celery blush! Observe how water is carried to the leaves along 'pipes' by carefully scraping off the outside layer of celery to reveal stripes. Trace the veins on leaves. Cut across the stalk to see a cross-section of the tubes. Does this work with celery with no leaves? What happens to the colour of leaves when they are deprived of light?
- Where does the water go (it is 'breathed out' by the leaves)? Test by tying a plastic bag around a growing leaf. Examine at regular intervals.
- Falling leaves: observe falling leaves in autumn. Describe the path of a falling leaf. Devise a fair test to find out which kind of leaf falls the fastest. Do most leaves fall on a windy day? Do they all fall at the same time? Collect falling leaves and sort them according to colour, size and shape.
- Take a leaf from one tree at regular intervals to show the gradual colour change: glue the leaves onto a chart or make a zig-zag book. Go out and crunch the fallen leaves underfoot: why do they make this noise? Does it work on a wet day?
- **In spring examine the buds opening on twigs.** Which buds open first? Watch leaves unfold and record the changes seen by close observational drawings. Look for leaf scars and growth scars on the twigs. Look for new shoots on evergreen trees and shrubs. In the garden observe the new growth of perennial plants.
- Investigate minibeasts and their relationship with leaves. Study the leaf-eaters (e.g. slugs, snails, caterpillars) and the leaf inhabitants, e.g. leaf-miners. Look for minibeasts in leaf-litter, e.g. woodlice, millipedes and spiders.

Leaves

- In cooking, take a close look at leaves that we can eat. Play 'pass the lettuce' and unwrap a layer each! Look at the pattern of veins on the back of each leaf. Order leaves according to size. Cut a red cabbage in half and look at the arrangement of leaves. Add water (try hot and cold) and observe colour change. Add vinegar and observe colour change again.
- Investigate the difference between perennial plants and annuals.
- Investigate pollution and dirt found on leaves.

Creative Activities:

- Leaf rubbings. Place a thin piece of paper over a leaf or an arrangement of leaves (use the under-side for best results) and rub with the side of a wax crayon. Make sure the paper doesn't move. Rubbings made in this way can be cut out and pasted onto a leaf collage.
- Leaf printing. Cover the reverse side of a leaf with thick powder paint. Gently place the leaf, painted side down, on to a sheet of paper. Cover with newspaper and gently press or roller. Younger children may find it easier to paint a leaf and place the sheet of paper on top. Rub gently with a roller to obtain the print. Use the leaf prints as a basis for pictures or patterns.
- **Make a 'falling leaves' display.**

  Thread leaves onto cotton and hang from display table, as shown.

  Sponge print a large tree. Display mounted prints and rubbings. Collect conkers, fruit, acorns, etc.

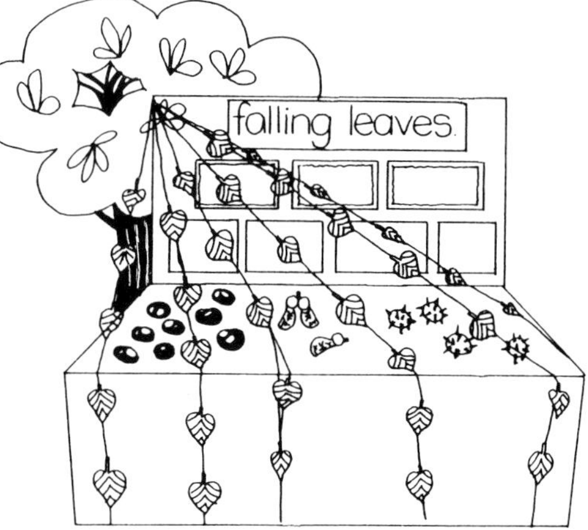

- 'In the Woods' picture. Paint outline shapes of a woodland scene with a mixture of thick black powder paint and PVA glue. Leave outline to dry. Now colour in the empty spaces using different shades of green (see photograph on p.19).
- Cut leaf shapes out of card. Coat the leaf shape with printing ink and gently place the shape onto the backing paper. Put a sheet of paper over the top and roller (see photograph above).

Seeds and Seedlings

Starting Points/ Discussion: Start with seeds with which the children are familiar (collected from their food): apple, orange, lemon, grape pips, plum, cherry and peach stones. What would happen if we planted some of these pips and stones? Look at other seeds. Group them into: seeds from fruit, seeds we eat, seeds for garden vegetables and flowers, tree seeds. Why do plants make seeds? Read *The Tiny Seed* by Eric Carle, Picture Knights.

⚠ **Science Activities:** **Some fruits and seeds are poisonous, e.g. laburnum, and some packeted seeds are treated with fungicide.**

Do all seeds that are the same grow into plants that are the same? What special part of the plant makes the seeds? Look at seed heads which have obviously come from flowers: dog rose, antirrhinum pods, dandelion heads. What do seeds need before they begin to grow? Sort the seeds by size. What is the largest seed you can find? The smallest seed? Why do plants produce so many seeds? Animal foods often contain seeds. Why do farmers or gardeners make scarecrows? What information is given on seed packets?

Vocabulary: Flower, fruit, seed, seed dispersal, pip, stone, germinate, roots, shoots, embryo plant, leaves, root hairs, light, warmth, air, water, soil, planting, sowing.

Collections: Seeds and seed cases; fruits: familiar ones and more exotic, e.g. kiwi fruit, star fruit, watermelon, etc. — and tomato, marrow, fruit sliced in different ways to expose the seeds; seed cases from flowers, e.g. poppy heads, love-in-the-mist, honesty (display these with a sample of the seeds); tiny seeds sprinkled onto adhesive tape to aid observation and handling; seed pods — peas and runner beans; conkers and their prickly seed cases (seasonal but always fascinating), along with sweet chestnuts, hazel nuts, beech nuts, coconut, etc.; dried grasses, ears of wheat or barley; from the kitchen: dried beans, rice, nuts and seeds used for flavouring, e.g. fennel, cardamom, caraway, etc.; seed jewellery and musical instruments using seeds; advertising posters or free samples from seed merchants; packets of seeds.

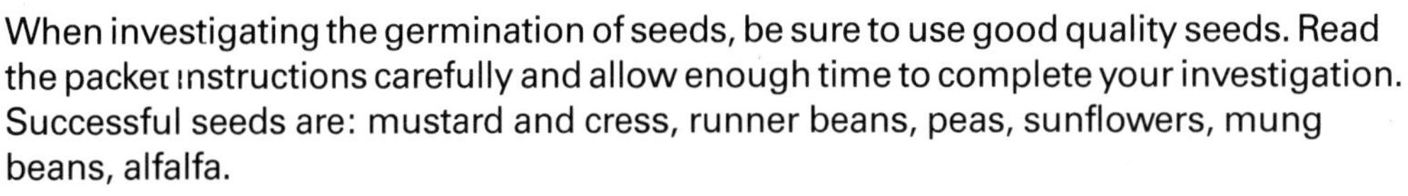

When investigating the germination of seeds, be sure to use good quality seeds. Read the packet instructions carefully and allow enough time to complete your investigation. Successful seeds are: mustard and cress, runner beans, peas, sunflowers, mung beans, alfalfa.

Seeds can be grown in a variety of containers — specialist seed trays are not necessary, although holes will have to be made in any containers in which you intend to grow plants. Garden soil can be used successfully if compost proves too costly.

⚠ **Hands must be washed after handling garden soil.**

- How can we make a seed grow? What will it need? A good starting point is — 'Why don't seeds grow in the packet?'.
- Try growing cress seeds on saucers lined with cotton wool — water one and leave the other dry. Place them on a window sill or radiator and observe what happens. Repeat this, water both, but put one in the fridge. Repeat again, water both saucers, but put one in a plastic bag and squeeze as much of the air out as possible. Secure tightly. Can the children now say why seeds don't germinate in the packet? What would happen if the packet got wet? Do soaked seeds germinate more quickly than dry seeds?
- What is inside a seed? Investigate using soaked seeds or fresh peanuts.
- Keep a seed diary. **Grow a runner bean in a wide-necked jam jar with rolled sugar paper and damp sand.** Position the bean between the paper and the inside of the jar. Keep the sand moist. Use a soaked seed. Observe carefully every day. Draw a picture — use a strip slider for recording (see strip slider — IN THE GARDEN).

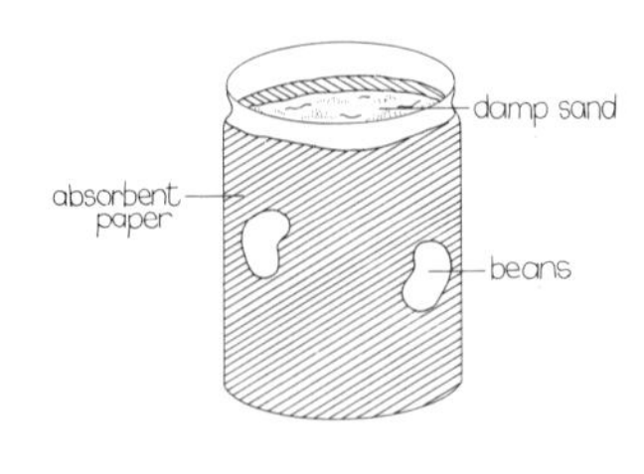

Seeds and Seedlings

- Do seedlings need light to grow? Use a shallow container — line with damp paper towels and sow cress seeds evenly on the paper. Place on a window sill, keep moist. When the seeds have germinated, **place a cone of black paper over approximately half the container.**

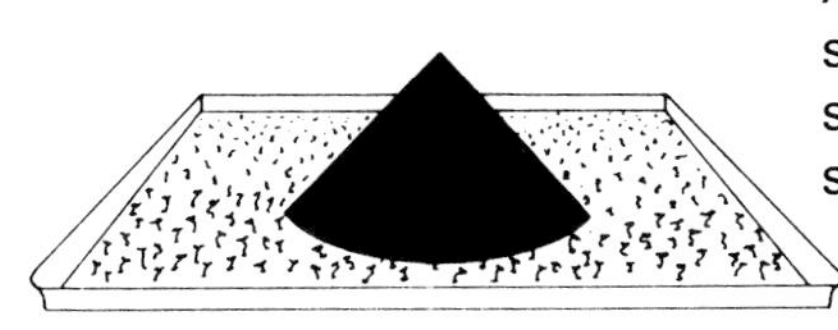

After a few days, or when the uncovered seedlings are sufficiently grown, remove the black cone. What do the seedlings look like compared to the uncovered seedlings?

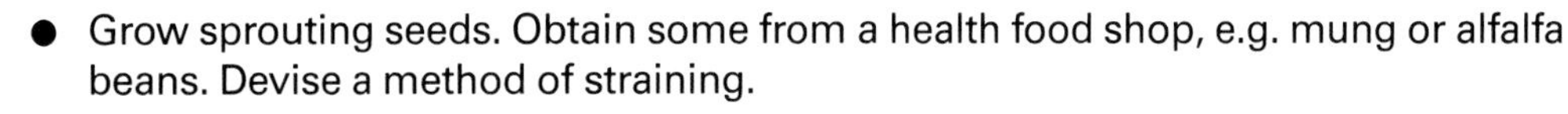

- Grow sprouting seeds. Obtain some from a health food shop, e.g. mung or alfalfa beans. Devise a method of straining.
- Investigate seed dispersal: those which use the wind to spread the seeds — dandelions; exploding pods — gorse; those which have hooks or barbs — what could these seed cases get caught on (animal fur and our clothes)?; those which fall — sycamore, ash; bird and animal dispersal through eating; water and sea dispersal.

Creative Activities:

- Slice open unusual or colourful fruits — observe using hand lenses and draw or paint (see photograph).
- **Make a scarecrow** to scare away the birds from seeds you may have planted outside. Make miniature scarecrows and stick them into flower pots filled with soil.
- Design and make a musical seed shaker.
- Observational drawings or paintings of seed heads, grasses, dandelions, etc. (White crayon on black, or white wax with a dark wash is particularly effective.)
- Make a seed picture frame. Arrange seeds in a pattern against a frame of card and attach using good quality glue. When dry, apply a coat of paint which has PVA glue mixed with it. Use it to display artwork or photographs.
- **Design a seed packet,** or container, perhaps for sale at a summer fair or Christmas market.

The Pond

Starting points/ Discussion: Spring and summer are the best times for studying ponds. Locate a nearby pond — it may be in the school grounds or in a garden or park or in the countryside. Obtain permission to visit the pond, especially if it is used by fishermen. A temporary pond can be created in the school grounds using an old plastic baby's bath or similar. Obtain some pond water and pond weed. Observe carefully and see what comes to live in the pond. Discuss the creatures that live in or near ponds. Have the children ever seen frogs or dragonflies? Discuss life cycles of these creatures. Read *Fish is Fish* by Leo Lionni, Picture Puffin. ⚠ **When visiting a pond it will be necessary to discuss safety aspects beforehand, e.g. slippery banks.** Discuss with the children the equipment they will need to take with them — jars and lids, icecream containers or other light coloured shallow trays, nets (including nylon sieves), pond viewers (see text), hand lenses, plastic bags, sticky labels.

Discuss the misuse of ponds e.g. for dumping rubbish. Is there any evidence of pollution and waste materials in the pond you visit? What could be done about this? How 'healthy' do the children think the pond is. Why? ⚠ **Always wash hands after contact with pond water.**

Vocabulary: Frog, toad, spawn, tadpole, permanent pond, temporary pond, algae, evaporate, hibernate, larva(e), plant eaters, predators, amphibian, snails, nymphs, pond skaters, whirligig beetles, newts, great diving beetle, water boatmen, dragonflies, water spider.

Science Activities:

- If visiting a permanent pond there will be many things to see before actually dipping in the pond. Look for water birds. (See AT THE PARK.) Look for holes in the banks which could be animal homes; look for tracks in the mud — can these tracks be identified? Look at the places where plants grow. Are there plants growing around the edges? Are there plants growing in the middle? Are some plants floating? Are some plants attached at the bottom of the pond by roots? Draw pictures of the plants and say from which part of the pond they come.
- Look at the surface of the water. Can any creatures be seen? Look at the way these creatures move on the water. Look for the stretched skin of the water around their feet. Why don't the creatures sink?
- **Make pond viewers and look into the pond.** What can you see? Make sure your shadow doesn't fall over the viewer.
  - Cut down plastic bottle, be careful of sharp edges.
  - Use strong plastic secured with elastic bands.

  ⚠ **Care is needed when using these to avoid falling in the water.** Take something waterproof on which to sit.
- Collecting pond creatures. It is important to stress that all creatures must be handled with care. If specimens are to be taken away for studying, make sure that there is adequate room for them in the containers and keep the containers away from direct sunlight. Separate creatures to avoid anything being eaten before you've had time to study it. *Return the creatures to the pond as soon as possible.*
  Sweep net through the water, wash net out into a shallow container filled with pond water. Collect some pond weed and put this into the water with the specimens. Identifying pond creatures can often be difficult owing to the different stages in life cycles likely to be found. Simple books will allow identification sufficient for the children's needs. Alternatively, sort the creatures into categories: plant eaters, predators, beetles, bugs, creatures with shells and so on.
- Make a mini-pond. Discuss ways that this could be done — old buckets, hollows in the ground lined with plastic, etc. Mosquitoes or gnats may lay their eggs; watch out for egg rafts. Observe the stages in the life cycle.

The Pond

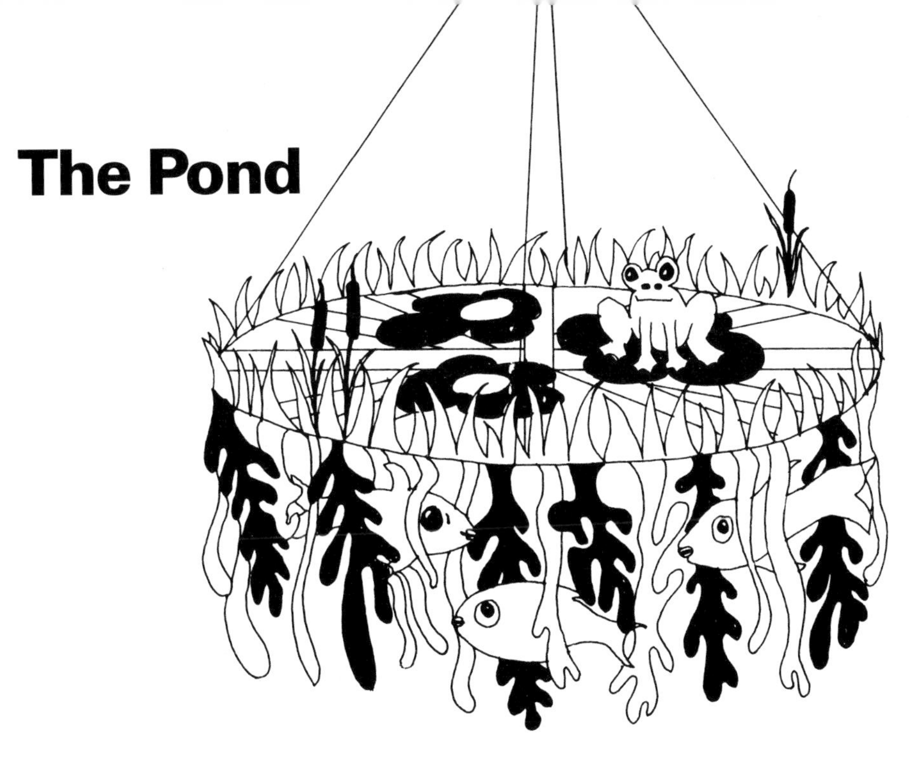

- Keep frogs' or toads' spawn. Remember to have only 10-12 eggs. Put pond weed, stones and washed gravel into a tank. Use pond water. Watch the development from egg to tadpole. Observe feeding. Tadpoles eventually need animal food. They can be fed with a small piece of lean raw meat — but don't leave this in the water for long as it leads to pollution. Let the young frogs go eventually — take them to a suitable habitat. Discuss the life cycle of a frog or toad.

Creative Activities

- Pond mobile. Use a hoop, attach weed and creatures with thread. Hang it from the ceiling. (See line drawing above.)
- Close observational drawings of pond creatures. Use hand lenses and try to get the children to draw them as accurately as possible.
- **Add the drawings to a cross section of a pond** displayed on the wall. Cut the plants out of paper; try to make them look like the plants that were actually growing around the pond. Sponge print the pond in green/blue and the banks in brown/black. Add labels.

- **What did you see through the pond viewer?** Use paper plates, sponge paint in blues and greens. Cut out pieces of tissue/crêpe/sugar paper for the pond weed. Cut out a picture of the pond creature that has been drawn and painted. Attach this to the plate on a folded piece of card to give an idea of depth.
- **Make life cycle models of frogs, etc.** Use a painted paper plate. Bubble pack could be used for the frogs' spawn. Add circles of black felt to the top of each bubble. Tadpoles could be made from painted packaging materials, tails could be added that are made from black fabric. Frogs could be made from green plasticine. Arrange on a paper plate.
- Pondscapes. Bubble print the background using two colours. Add crêpe paper or Cellophane weed and painted or printed fish. (See front cover.)

Wooden Things

Starting Points/ Discussion: The starting point could be a collection of things made from different woods, e.g. a jig-saw puzzle, a wooden bowl, a spoon, a xylophone, a chopping board. What do all these things have in common? How are they different? (Consider use, colour, rough/smooth, heavy/light, shiny/dull, painted/unpainted.) Observe patterns and markings. Do they feel warm or cold? Look at how wood has been used within the school, e.g. for flooring, window frames, doors, panelling, cupboards and furniture, rulers, pencils, toys, musical instruments, P.E. apparatus, fencing etc.

Investigation of wooden things could also follow on from:
— the study of trees in the environment
— the study of a building project taking place within the school grounds or nearby
— a comparison of different materials used in the kitchen (e.g. as conductors/non conductors of heat)
— making boats — making musical instruments.

⚠ Visit: a museum to look at wooden artefacts; **a local carpenter or timber yard; a building site.**

Vocabulary: Pine, oak, ash, walnut, cherry, mahogany, teak, maple, beech, balsa, veneer, plywood, blockboard, chipboard, hardboard, pulp, sawdust, wood shavings, splinters, planks, logs, trunk, bark, sapwood, heartwood, grain, knot, growth rings, seasoned, rot, woodworm, charcoal, carpenter, joiner, timber yard, paper, forest, Forestry Commission.

Collections: Wooden things, cooking utensils, bowls, egg cups, spoons, boxes, toys (including jig-saw puzzles, puppets, games); musical instruments, beads, clogs, picture frames, models, samples of wood (ask your local secondary school, a carpenter, or joiner for different types of wood of similar dimensions); types of board, e.g. chipboard, blockboard, hardboard, logs, wood shavings, sawdust, matchsticks, lollysticks, twigs, pieces of bark; damaged wood, e.g. with woodworm holes, rot or very knotty wood, charred wood; charcoal; polished, varnished, painted, decorated items as well.

⚠ **Carpentry tools: stress safety in use of tools and adhesives.** Pictures of trees, modern and antique wooden furniture.

Science Activities:

- 'Adopt a log' — observe a log outside over a period of time. What is bark? What does it do (protects and insulates, prevents water loss)? Look for growth rings, sapwood, heartwood. Look for minibeasts, rotting wood, moss and fungus.
- ⚠ **Burning wood.** What happens to wood when it burns? Observe a burning match. How does it change? Can you draw with the charred wood? Show children charcoal and let them compare: feel how light it is, how easily it is broken.
- Investigate: why do we often have wooden handles on saucepans? Why do we use wooden spoons and spatulas for cooking? ⚠ **Put a wooden spoon and a metal spoon on a radiator or into a container of hot water.** Which feels hotter after a short length of time?
- Investigate: does electricity travel through wood? Make a simple circuit (see BATTERIES AND BULBS) and test wood samples.
- Investigate: do magnets work through wood? Does the thickness of the wood matter? Vary the type of wood used.
- Make sounds with wood: compare sounds made with different lengths of wood; strike wood with wood, wood with metal; compare hollow and solid wood; compare wet and dry wood; does it matter where the wood is resting?
- Does wood float (some woods are very dense and do not float)? Test your samples of wood to find 'good' floaters. Does the shape/thickness of the piece of wood matter? Does the wood change colour? What happens if you leave the wood in the water for a long time? Compare weight before and after soaking.

Wooden Things

- Carpentry. Investigate ways of joining wood together, e.g. glues, nails, screws. Look at joints used in furniture. Investigate cutting and sawing: consider the effects of grain. Carry out strength tests: does wood bend? Look at the way man-made boards are constructed for strength and rigidity. Look at the way shelves and bridges are supported. Glue together lollysticks and matchsticks for strength. Smoothing and sandpapering: what happens when you rub wood with sandpaper? What happens to the sandpaper? How can you test the hardness of woods?

Creative Activities:

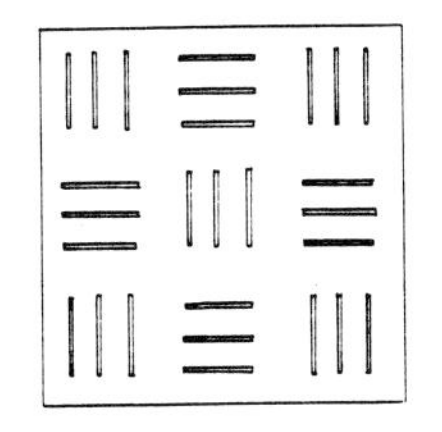

- Rubbings. Use collected wood samples, bark of trees, wood found in classrooms and playground (wooden benches, fencing, gates, climbing frames, etc.). Go on a 'knot hunt' outside. Use brown crayons on thin white paper for best results. Encourage the children to cover the whole piece of paper to experiment with patterns.
- Wood collage. Stick wood shavings, sawdust, matchsticks, lollysticks, twigs and pieces of bark, off-cuts of wood, on to cardboard or paper. Cover large areas with glue and sprinkle or place wood to make a picture or pattern.
- Use charcoal to make close observational drawings of twigs, branches, bark and wood patterns.
- **Make peg dolls and wooden spoon dolls;** make matchstick or lollystick models; make boats; make a chassis for a moving toy.
- Stick-printing patterns. Print with ends of wood pieces, beads, rulers, etc.
- **Make a printing block with matchsticks stuck on to a wooden block.**

In the Garden

Starting Points/ Discussion: Visit a garden and a garden centre. What do we find in gardens? What do we do in gardens (play, relax, barbecue, gardening, bonfires, dry the washing, etc.)? How are gardens different at different times of the year? When would we find most things growing? When do we spend most time in the garden? Who visits our garden (friends, refuse collectors, birds, foxes, hedgehogs, cats)? What sounds can we hear in the garden (insects, birds, lawnmowers, people talking, hoses spraying water)? Read 'The Garden' from *Frog and Toad Together* by Arnold Lobel, A Young Puffin.

Vocabulary: Plants, trees, bushes, flowers, vegetables, ponds, patios, hedges, walls, fences, sheds, greenhouses, garden furniture, soil, stones, minibeasts, growth, gardening, weeding, spade, trowel, fork, rake, sieve, grass, lawn, mowing, watering, barbecue, pollen, sepals, petals, stamens, carpels, nectar.

Collections: Bird baths, birds' nests, bird tables, pictures of common birds, grass seed planted and growing in pots/trays, packets of flower and vegetable seeds, grow bags planted with tomatoes, beans, etc.; flowers and plants growing in pots, bulbs planted in pots; garden tools — spades, rakes, trowels, sieves, etc. (nothing sharp or dangerous); watering can and rose; gnomes and small garden statues; flower press, dried flowers, pictures of gardens and garden furniture from catalogues, etc.; small pieces of garden furniture.

Science Activities: Gardens lend themselves well to seasonal activities and can be used in a wider context, e.g. in a topic on homes or summer time.

- In the spring leave out a selection of materials that birds may use for nest building. Look at old nests and decide which materials would be suitable. How will you know if the birds have taken them or whether the wind blew them away?
- Study birds: why do they make nests? Look for signs of broken egg shells on the ground. What do baby birds look like when they are newly hatched? How do they feed?
  NB: emphasise that nests or young birds must not be touched in the spring time.
- Look for signs of new growth on the trees in the garden. Observe buds and catkins including pussy willows. Obtain some branches and watch the buds open inside the classroom. Why do these buds open more quickly?
- **Grow bulbs in pots in the classroom** — tulips, daffodils, amaryllis, etc. Make careful measurements of the growth of an amaryllis. At what stage does it grow most quickly/slowly? Observe the flowers. What can you see inside? (See also SEEDS AND SEEDLINGS.)

Always wash hands and scrub nails after handling soil.

- **What is in the soil?** Obtain some garden soil. Look at it carefully with a hand lens. Does it contain stones, broken pottery, small pieces of plants, minibeasts? Shake with water in a large jam jar and allow to settle.
- What lives in the soil? Dig a hole in the ground. What can you find? Which minibeasts like damp dark places? (See MINIBEASTS.) Make a wormery.
- Which soil lets water pass through? Have a race — each child chooses a different kind of soil. Put equal amounts of soil inside a coffee filter paper and place over a jar. Add the same amount of water at the same time. Which one do they think will let the water through the quickest — why?
- What happens if we cover grass? Cover a small area of grass with an old upturned flowerpot, an old tin etc. Observe daily what happens to the grass under the cover. Why does this happen?

In the Garden

- What is inside a flower? Choose big flowers such as tulips, nasturtiums, poppies or dog roses to study. Carefully pull the flower apart and look at all the different parts. Look for pollen and nectar. Does the flower have a scent? Cut through a fertilised flower and show the new seeds growing. Explain that these will grow into new plants of the same kind. Attach the parts of the flower to a card with adhesive tape and label where applicable.

Creative Activities:

- Plasticine and sponge prints. Shape the plasticine (or clay) into petal shapes and leaf shapes, and use these for printing. Details can be added by marking with a pencil onto the plasticine. Use sponges to absorb excess paint from the blocks. This technique is particularly effective for making pictures of flowers that grow closely together — daisies, buttercups, forget-me-nots. Use toned paper for a softer effect.
- Pressed flowers. Collect suitable flowers and press them either by using a flower press or heavy books. Glue onto paper and use for greetings cards, gift tags etc.
- Paint flowers using painting inks (see photograph above).
- Make a seed tray garden (grass seed is very effective). Use mirrors for ponds, add a rockery, make a path using gravel and add small plants. Make models of garden furniture.
- Garden layer pictures. Begin by sponging a bright background. Use collage materials to make flowers, trees and minibeasts (see photograph above).
- **Make a season chart for the garden,** showing the same scene during each of the four seasons.

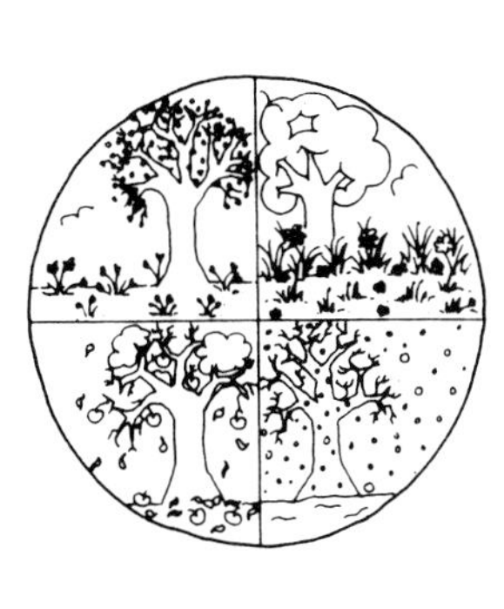

At the Park

Starting Points/ Discussion: Begin by visiting a park or play area with a good selection of play apparatus. Which rides go round and round, up and down, backwards and forwards? Which one is slippery? Which rides need a push to get going? What are the rides or climbing frames made from? How are they attached to the ground? What kind of surface is under the play apparatus — is it grass, concrete, rubber, tarmac, filled with bark? Is it a safety surface? How can you tell?

What other things could we find in a park? Grassy areas, flowerbeds, trees and bushes, benches, ponds or lakes, birds, squirrels, icecream shop or café, tennis courts, bowling greens etc.

What can we do in the park? Play games, fly kites, play with frisbees or balls, have picnics, feed the ducks, collect leaves and tree seeds, sail boats on the lake. How will we know what to wear to the park? Listen to or watch weather forecasts. Is there a lot of litter in the park? Where is there most litter? Are there enough bins?

Who looks after the park? Invite the park keeper in to talk to the class or arrange to meet him or her there. Read *A Walk in the Park* by Anthony Browne, Picturemac.

Vocabulary: See-saw, swing, roundabout, climbing frame, slide, safety surface, weather forecast, picnic, park keeper, moorhen, swan (cygnet), heron, mallard (duck and drake), Canada goose, great-crested grebe, tufted duck, swim, dive, preen.

Collections: Leaves, berries, seeds, etc. from trees in the park; pictures of trees according to time of year that park study takes place — compare with other seasonal pictures; pictures of squirrels and water birds with names displayed; a picnic box and a pretend picnic of foods that would be suitable to take; balls, kites etc.

Science Activities:

- After playing on the swings discuss how the children made themselves swing. Did someone give them a push? Did they swing their legs? **Make some model swings in the classroom.** One way would be to use strings attached to a short plastic ruler or a piece of stiff card (see line drawing).

  A swing is a pendulum — where else are pendulums used?

  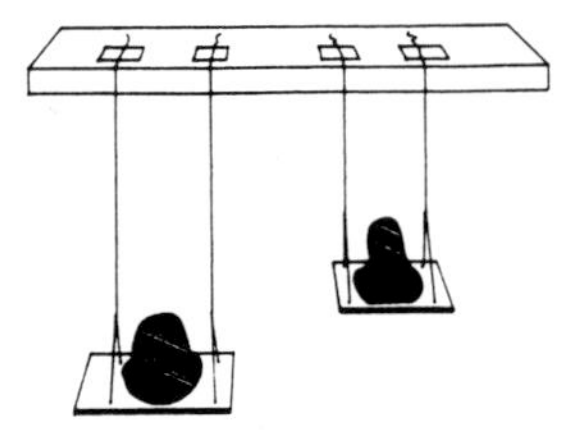

  Investigate whether the swing is faster or slower when the length of strings are varied (made shorter or longer). Add weights (plasticine balls) to the swings — does this make any difference?

- How do see-saws work? Whilst at the park investigate under supervision how a see-saw works. What happens if an adult sits on one end and a child on the other? Can you make the see-saw balance? Where is the see-saw supported? Are both arms the same length? In the classroom make model see-saws. A ruler could be used with a triangular prism shape (Toblerone box or wooden 3D shape) for the pivot. Investigate what happens if small objects are placed on each end — can you make the see-saw balance? Make plasticine models of people and put them on the see-saw — make one bigger than the other, what can be done so that the see-saw will balance?

- Does it make any difference what clothes we wear for sliding? Did some children slide down the slide more quickly than others? Why? What happens if you put your feet down whilst sliding? Is it easy or hard to go down a slide with shorts on? What did the surface of the slide feel like (smooth, shiny, slippery)?

At the Park

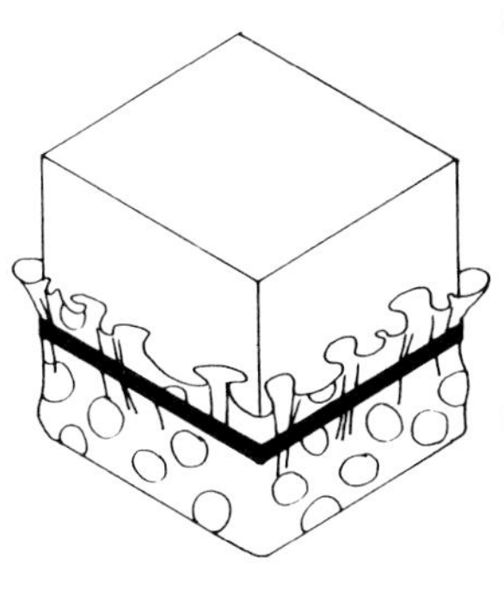

- Obtain a selection of fabrics, and sort into good sliders and bad sliders. One way to investigate would be to attach pieces of the fabric to building blocks with elastic bands (try to include fabrics like leather or PVC cloth etc.). **Make a model slide using an old tin tray.** Decide upon a suitable slope. Place fabric covered blocks at the top and let go, all at the same time. Which one was the fastest/slowest? Repeat several times to ensure fairness. Discuss why some fabrics are more 'slippery' than others.
- Observing water-birds. Discuss the names of the birds likely to be found on the park lake. Use pictures to help with identification. Watch the birds feeding. Which ones dive, up-end in deep/shallow water, which birds graze on the grass? Look at the birds' feet — which ones have webbed feet? How do webbed feet help with swimming? Observe the birds' bills. Which ones are long and pointed, which ones are wide? What do the birds eat and how are their bills suited to this?

Creative Activities:

- Sponge paint a scene from the park (see photograph above).
- Make a model lake. Use Cellophane or foil for the lake. Add stones etc. around the edges. Make models of water birds out of clay or plasticine.
- Make collages of different birds found on the pond. Cut feather shapes from tissue or crêpe paper. Swan feathers could be made by cutting around the shape of hands on white tissue paper. Overlap them and stick them onto the swan shape.
- Collect leaves from the trees in the park. Make leaf rubbings. Make bark rubbings of the trees and display with the appropriate leaf rubbing and label.
- Design a poster to encourage people to keep the park clean and litter free.
- Design a park. Where would you site the different pieces of play apparatus? What kind of surfaces would you use?
- Make a model of your ideal park using waste and scrap materials. Can you make your apparatus move?

Minibeasts

Starting Points/ Discussion: The topic of minibeasts could form part of a study of the garden, the pond, a study of eggs, leaves, change, or movement. The starting point may be a minibeast brought into school by a child. To some extent the work will depend on the time of year e.g. in autumn leaf litter and ripe fruit could be examined; in early spring the children could look for pupae and cocoons; in early summer caterpillars and aphids will be much in evidence; later on in the summer term flying insects, ants, ladybirds and spiders can be observed. There are many story starting points e.g. *Charlotte's Web* by E.B. White, Puffin, and the imaginative *If At First You Do Not See* by Ruth Brown, Beaver Books. For the most part discussions will be determined by the children's observations of minibeasts in their natural habitats, but a visit from a local expert such as a bee-keeper or a gardener interested in using 'organic' methods of growing and pest control will add much to the development of the topic.

Vocabulary: Minibeast, larva, pupa, cocoon, eggs, chrysalis, life cycle, animals without backbones (invertebrates), insects, symmetry, respiration, feelers (antennae), wings, shell, proboscis, spinneret, head, thorax, abdomen, webs, nest, habitat, beehive, leaf litter, ants, bees, butterflies, caterpillars, flies, ladybirds, millipedes, silkworms, slugs, snails, spiders, woodlice, worms.

Collections: Pictures of insects and minibeasts; leaves, bark, flowers, fruits and seeds that show insect damage; dead insects, honeycomb, old wasps' nest, eggs, pupae. Talk about methods of collecting live minibeasts for short term observations in the classroom. How can you lift creatures without damaging them? Discuss conditions the minibeasts are going to need during their short stay (food, soil, water, air, shade or shelter, etc.).

⚠ Replace stones and bark after observations. **COLLECT WITH CARE!** Provide hand lenses.

Science Activities:

- Go on a minibeast hunt. Can the children suggest where they might find minibeasts? Test out their predictions. Try the following places: in grass, under trees, on trees, leaves and flowers, near and on a wall, under stones, under a log, in a hedge, on or in water, in soil, on vegetables and under a flowerpot. What can you learn about minibeasts by looking at their habitats? Visit the habitat at different times of the day and in different weather conditions, e.g. why do many creatures like hiding under stones (cool, dark, damp, safe)? Can you identify the minibeasts you have found? How many different kinds did you find in one place? Do some creatures appear in lots of places? Or in only one place? Draw a map, a picture or a diagram of the area studied. Put 'flags' or markers on your map for each kind of creature found. Are the creatures easy to see in their habitat, or are they effectively camouflaged?
- Look at minibeasts carefully. Observe colours and patterns. Are they of more than one colour? Are they symmetrical in shape? Can you devise a method of measuring your minibeasts? What problems do you encounter in measuring your creature? Do all creatures of the same kind look identical (i.e. do they all have the same markings or patterns)? Look at the shape of the head. Count any legs. How many wings? How many body parts or segments? Do they have antennae, a shell, etc.? Can you see a respiration hole? Can you see their eyes? Play a game of 'What am I?' in which a creature is identified by its attributes.
- How do the minibeasts move? Movement can be observed both in the classroom and in the creature's natural environment. Compare the speed and distance that different minibeasts achieve. Devise tests (taking great care not to harm the animals) to see how their movement changes over different terrains, or when faced with obstacles. Are there any specific conditions that encourage the creature to move? (e.g. temperature, light/dark, wet/dry). Can you see any movement while the creature is stationary?

Minibeasts

- Study how the minibeasts feed. How can you find out what the creature prefers to eat? How do they capture their food? Discuss how some minibeasts can be regarded as friends to the gardener or grower by helping to control the population of pests.
- Study the life-cycle of different minibeasts. Can the children find any similarities between different creatures? Are there any recurring patterns?
  (e.g. eggs → larvae → pupae → adults or
  eggs → small adults → adults)
  Children may become interested in the number of eggs laid, where they are laid, whether the adult cares for its young, whether there is a good chance of survival or not.
- **Spiders**
  Look for spiders out of doors. Search likely places for webs. Look at the shape of the web and describe it. Sketch it and look up its name in a reference book (e.g. orb or hammock web). Observe a spider making a web. What happens when you blow gently on the web, or touch it? Can you see any food in the web? Look for different kinds of spiders and compare them e.g. colour, size, shape of body, hairy/non-hairy, shape of legs, etc. Look for spiders' eggs in autumn, baby spiders in spring.
- **Caterpillars, Butterflies and Moths**
  Look for caterpillars, or eggs on plants. They can be brought into the classroom as long as there is a readily available supply of leaves for the caterpillars to feed on (the leaves that they were found on). Observe how the caterpillar eats. Where does it start on the leaf? Does it always eat in the same direction? Changes can be observed as the caterpillars feed and grow. How does the caterpillar's skin split? Watch as the pupa develops. What does it feel like? How is it different from the caterpillar? Observe the emerging butterfly or moth. Set it free as soon as possible after observations. Find out about butterflies and moths and camouflage (e.g. the eyed hawk moth).

Minibeasts

- **Snails**
  Look for snails in likely places e.g. where it is damp and dark, on foliage, under garden rubbish, etc. Follow their trails early in the morning to discover where they have been feeding. Compare the shells of different snails. Are they all the same size, shape, colour? How many different kinds of slug and snail can you find? Look for eggs in the soil. Bring snails into the classroom. Remember they will need food and a damp habitat for their short stay. When it is hot and dry the snail will seal its opening. Can you provide the right conditions for it to re-emerge? Look for a snail's tentacles, eyes, respiration hole. Observe how it moves (place on a piece of glass or plastic and observe from underneath).
- **Ladybirds**
  Look for ladybirds and their larvae on roses and plants infested with aphids. Look at the ladybirds' wings. Their flying wings are underneath their coloured patterned front wings. Look at the arrangement and number of spots. Do all ladybirds have the same markings? Look for eggs, the growing larvae and the pupae. Ladybirds hibernate in winter, crawling into cracks in pathways and the bark of trees, and appear to be dead.
- **Worms**
  Worms are so familiar to children that they may not have observed them very closely at all. Collect worms carefully from the school grounds. Keep in layers of dampened soil in a jar; take care to keep them dark by wrapping a collar of paper or card around it. Put pieces of food (vegetable matter, leaves) on top. Compare the worms. Are they all the same length? Are they all the same colour? Can you see segments, bristles? How do they move? Can you tell which end is which?
- **Ants**
  Ants are easy to find. Look between paving stones, by walls, and near and on rose bushes. They live and work together. Observe ants travelling outside. Are they all going in the same direction? How do they move? Are any of them carrying anything? Test to see what happens when you put something in their path. What kind of foods do they seem to like best? Find out about how they cooperate in the nest. Look for the young queens and male ants flying on a warm day in summer.
- Other minibeasts to look out for include centipedes, millipedes, woodlice, grasshoppers, crane-flies (daddy-long-legs), wasps, bees, and earwigs.

Creative Activities:

- Make close observational drawings and paintings of minibeasts at different stages of their development using a variety of media including pastels, charcoal, inks and felt-tipped pens. Keep an illustrated diary of any changes observed.

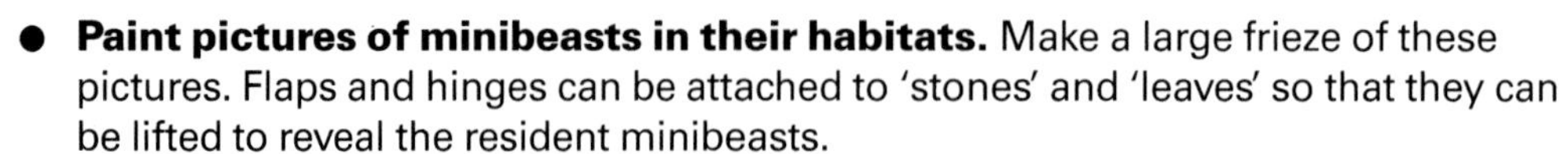

- **Paint pictures of minibeasts in their habitats.** Make a large frieze of these pictures. Flaps and hinges can be attached to 'stones' and 'leaves' so that they can be lifted to reveal the resident minibeasts.
- Draw or paint flying minibeasts and cut them out. Attach to paintings of their surroundings (e.g. a flowerbed) by a piece of curled card — so that the creatures hover and move with the breeze (see photograph on p.31).
- Investigate the patterns of minibeasts e.g. spirals, spots, stripes, patches and delicate wing patterns. Make printing blocks of these patterns with card and string or polystyrene 'pressprints'. Make camouflage pictures of the creatures against their backgrounds.
- Make an 'ant-trail' frieze. Cut out ant shapes from black sugar paper (these can be cut on the fold to produce symmetrical figures). Paste onto a printed background arranged so that the trail of ants follows a winding pathway.

Minibeasts

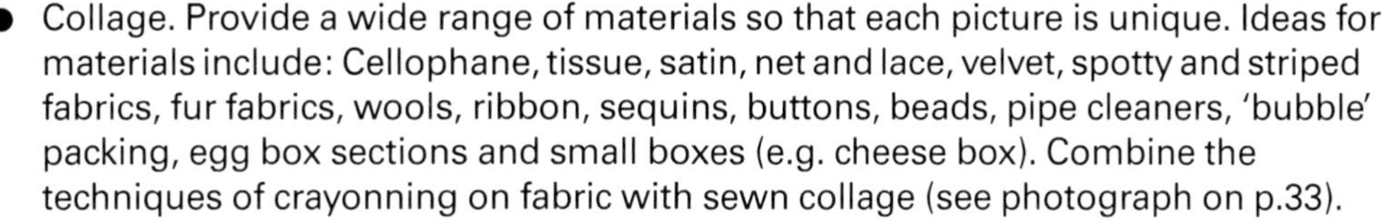

- Collage. Provide a wide range of materials so that each picture is unique. Ideas for materials include: Cellophane, tissue, satin, net and lace, velvet, spotty and striped fabrics, fur fabrics, wools, ribbon, sequins, buttons, beads, pipe cleaners, 'bubble' packing, egg box sections and small boxes (e.g. cheese box). Combine the techniques of crayonning on fabric with sewn collage (see photograph on p.33).

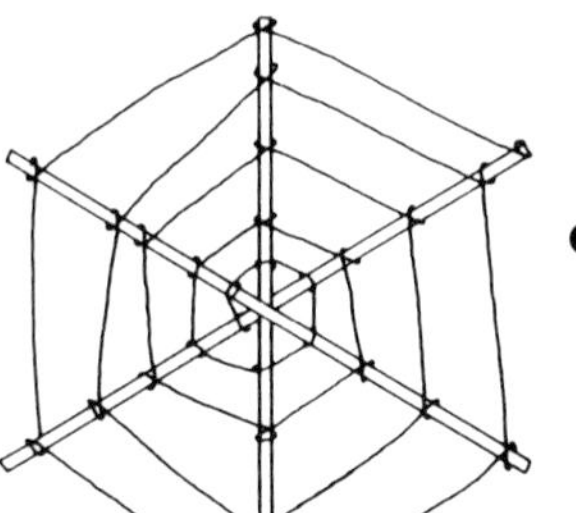

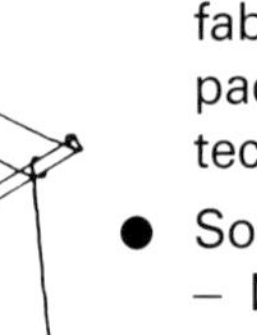

- Some ideas for spiders' webs.
  - Draw with silver or white crayons, pressing heavily, and add a wash of thin paint or ink. Draw the spider with pastel crayons.
  - Paint webs using gold or silver powder paints on black paper. Stick sequins onto the intersections of the web.
  - **Make webs by weaving silver and gold threads on sticks,** or weave onto bare twigs and branches, using grey, white, black, 'sparkling' wools.
  - Make a web-shaped sewing card. Older children will be able to design their own. Make a pipe cleaner spider to sit on the web.
- **Write poems or stories about a snail or a caterpillar on creature-shaped paper.** The story or text follows the shape of the creature and the pathway it has taken.

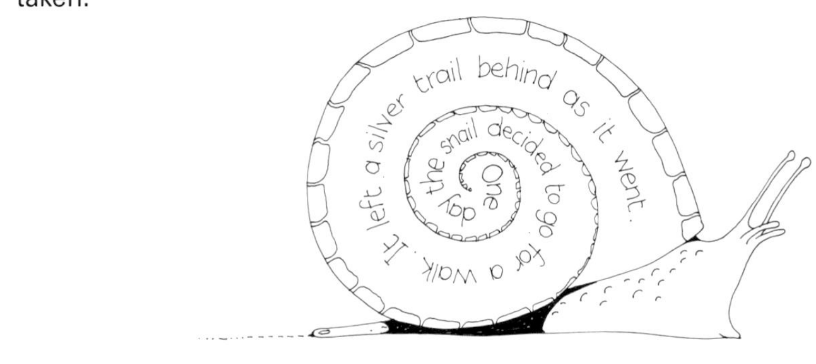

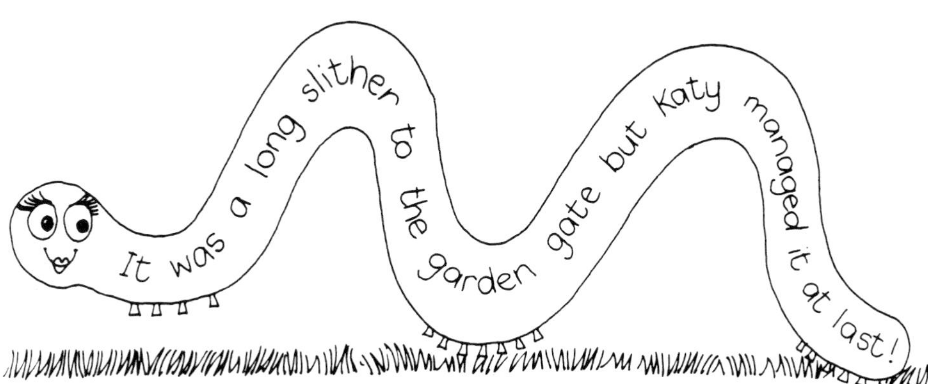

- Coil plaited paper or corrugated card into a spiral to make a snail's shell.
- Make models with clay, dough or plasticine. Paint dough models with PVA glue to make them shiny.
- Make built-up 3D collages of minibeasts. On a firm base of card stick crumpled balls of kitchen paper or newspaper into the rough shape of the creature's body. Paste layers of newspaper over the top to gain the shape wanted, finishing with a sheet of tissue. Paint and coat with PVA glue if a shiny effect is needed (e.g. for ladybirds, beetles, earwigs). Add wings, legs, antennae, and sponge paint or print the background.
- **Make a book to show the life-cycle of your minibeast.** Make a zig-zag book first – then join the first page to the last and paper-clip them together (like Victorian 'round' books). Hang books up.

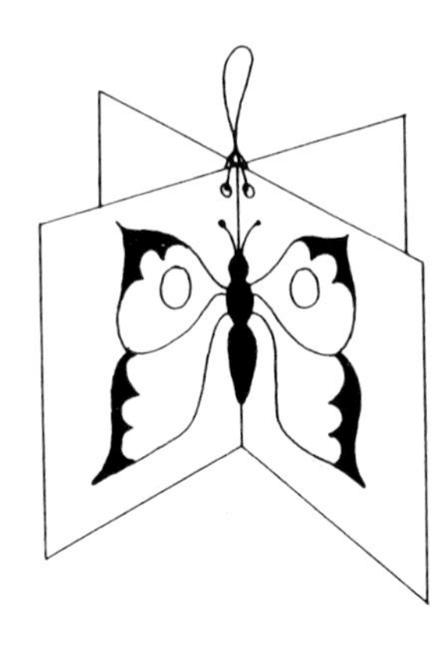

Bridges

Starting Points/ Discussion: Look at bridges in the locality. Who crossed a bridge on the way to school today? Why did you need to cross a bridge? What materials are used to build bridges, e.g. brick, concrete, wood, steel. How can you travel across the bridge? e.g. by car, bus, train, on foot. Talk about the historical importance of bridges; how towns and villages, for instance, grew up at important river crossing points. Older children might consider how you would choose the best site for a bridge across a river, e.g. the narrowest point of the river, type of soil, materials readily available, etc. Look at famous world bridges. Introduce the names of different types of bridge, e.g. suspension, arch, cantilever.

Vocabulary: Beam, cantilever, suspension, arch, swing bridge, footbridge, flyover, aqueduct, viaduct, pontoon, Bailey, humpback, clapper, breeches-buoy, drawbridge, abutments, keystone, span, causeway, stepping stones, ford, support, brackets, plank, cables, girder.

Collections: Pictures of bridges of all kinds; construction toys and materials for making model bridges: Lego, Meccano, wooden bricks, straws, string, rulers (both wooden and plastic), paper and card of different thicknesses, house bricks, pieces of balsa wood, plastic, 'junk' materials, e.g. polystyrene, weights, books, model cars and other vehicles.

⚠ **Make sure that experiments involving the children are carried out at low levels, over safe surfaces. Ensure that heavy weights do not fall on feet!**

Science Activities: Science investigations concerning bridges could arise from a story beginning, such as *The Lighthouse Keeper's Catastrophe* by Ronda and David Armitage, Picture Puffin — how can the gap between the lighthouse and the cottage be spanned? Or, in *The Cow who fell in the Canal* by Phyllis Krasilovsky, Picture Puffin — how many kinds of bridges can you see? How do they work? With very young children, a story such as *The Three Goats* might lead to a challenge — 'Make a bridge for the three goats to cross the river'. Specify the width of the river and emphasise that you cannot go into the river because of the troll.

- To show realistically an engineer's problem of spanning a gap, go out into the playground and pace out or measure the distance spanned by a local or a famous bridge. In P.E. make simple beam bridges with the apparatus for the children to cross. Emphasise the need for adequate supports at either end and look at the flexibility of wooden planks as weight is moved along them.
- Make model bridges using construction toys, e.g. Duplo, Lego or Multi-link cubes. Look at the need for support in the middle of the bridge and for weight at the ends of the structure.

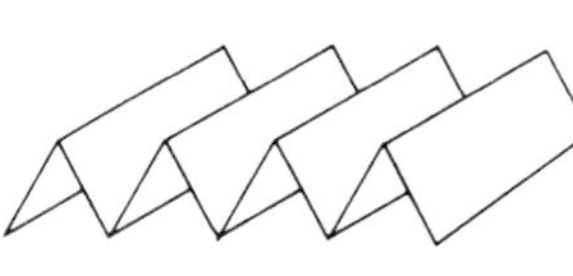

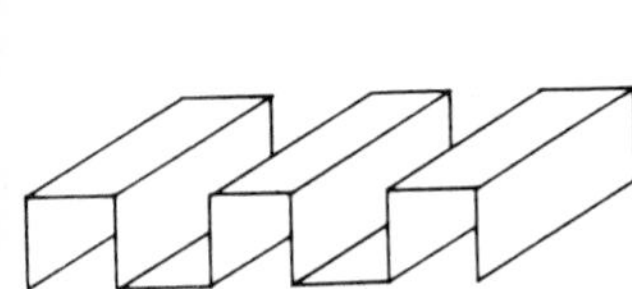

- Make simple beam bridges with pieces of card spanning two house bricks or piles of books. Put weight on the bridge. Look at the card — is it sagging? Increase the weight and look at what happens. How could you strengthen your bridge? **Try to make the card stronger by changing its shape.**
- **Make bridges with tubes of paper or card** or with bundles of straws. Try lollysticks, pieces of balsa wood, corrugated cardboard and plastic.

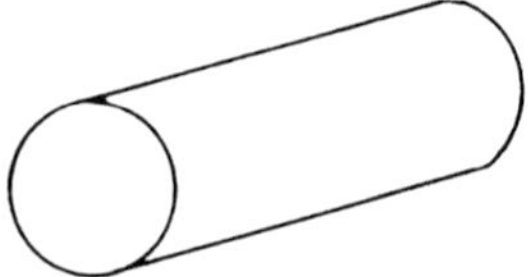

- Rope bridges sag a lot, but they are very strong. Why? Make a breeches-buoy with a rope and a curtain ring.
- Look for arch bridges in the locality. **Make an arch between two piles of books, with a piece of card.** Add extra support if needed. Look for the keystone at the top of an arch (look at porches and doorways).

Bridges

- Many bridges are cantilever bridges; they have arms projecting from their supports connected together in the middle of the span. Investigate the making of cantilever bridges with piles of books and a plank. How can you make it balance (by adjusting the weight of the abutments)?
- **Suspension bridges.** Hang string between two chairs. Cut equal lengths of string and tie to chair strings to form a cradle.

  Slide a piece of card or wood into the cradle. Press down on the plank. What happens to the support? How do they solve this problem with suspension bridges?.
- Investigate the mechanisms of moving bridges.

Creative Activities:

- Paint or draw bridges from observation.
- Paint bridges showing reflections in water.
- Make model bridges from junk materials. Can you make a moving bridge, e.g. a drawbridge? Devise a winding mechanism.
- Print using everyday objects — look at the patterns of stones on arched bridges and print with ends of sticks, Lego, beads, bricks, etc. to show the structure (see photograph above).
- Make collages of straight-line patterns found on bridges and their supports (using straws). Use string for suspension bridges.
- Paint mural of *The Three Billy Goats Gruff* or *The Lighthouse Keeper's Catastrophe.*

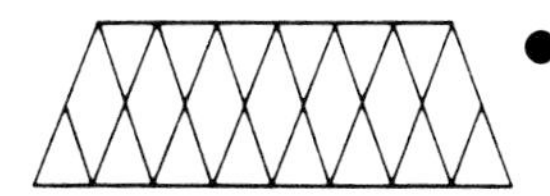

- **Look at the patterns and shapes found on iron and steel bridges** (especially the triangular patterns found in the framework of truss bridges). Recreate these patterns by printing with the side of a piece of card.

Metals and Magnets

Starting Points/ Discussion: The study of metals and magnets may arise from a general topic such as 'Toys'. A visit to a museum or art gallery might lead to the study of metal jewellery, coins, machinery, tools, armour, implements or sculpture. A good story beginning is *The Iron Man* by Ted Hughes, Faber and Faber. Begin the discussion by identifying metal objects belonging to the children, e.g. a watch, slides, buckles, zips, jewellery, badges, coins, scissors and toys. What does metal look like? Talk about the colours of different metals. Are they all shiny? What does metal feel like? Is it cold, smooth, sharp, heavy, hard, pointed? Introduce the names of common metals and discuss their uses. Talk about 'precious' metals.

⚠ Talk about safety. **Some metal objects might be rusty, sharp or poisonous.**

Vocabulary: Aluminium, brass, bronze, iron, copper, gold, lead, silver, tin, steel, nickel, platinum, zinc, alloy, corrosion, rust, enamel, hammering, tarnish, minerals, ores, mining, magnet, electromagnet, attract, repel, force, poles, north, south, compass, wire, rod, bar, girder, pipe, cable. Descriptive language: 'hard as iron', 'nerves of steel', 'like a lead balloon', 'like gold dust'.

Collections: Metal objects, including different forms of the same metal, e.g. foil, wire, piping, springs, sheets, rods and chains, (plated objects and precious metals could pay a brief visit to the classroom); collections of rocks, minerals and ores; magnets, magnetic games and small magnetic items: nuts, bolts, screws, fasteners; coins, cans, kitchen equipment, toys, tins, badges, etc., pictures of large machines, bridges, cars, trains, bicycles.

Science Activities:

- First sorting of objects into sets of metal objects/non-metal objects.
- Sort metal objects according to colour, non-shiny/shiny metals, heavy/light, painted metal/not painted.
- Sort metal objects according to: will/will not be attracted by a magnet. Display the sets. Record results by drawing charts.
- Investigate the power of magnets. Feel the forces of attraction and repulsion.
- Do magnets work through things? Try water, glass, plastic, wood, cardboard, paper, other metals. Does the thickness of the material matter, or the strength of the magnet? Devise fair tests.
- Can you make a magnet? What happens when you stroke a nail with a magnet (remember to stroke in the same direction)?
- How can you test the strength of a magnet?
- Can you find out how to make a simple compass?

- **Make sounds with metals.** Collect and play metal musical instruments, e.g. bells, chimes, chime-bars, cymbals, triangles, Jew's harp, gong, steel drums, metallophone. Look at musical instruments from different countries. Observe shape, thickness and length of pieces of metal, and the sounds they make. Make a pots and pans band! Investigate the sounds produced by metal hitting metal, wood hitting metal, etc.
- Look at reflections in shiny metal objects, e.g. a shiny kettle, spoons, foil.
- Do metal objects float? Does the shape/thickness make any difference?
- Which metals go rusty? What has happened to them? How can we protect metals, e.g. by painting, oiling, plating, varnishing?
- ⚠ **Investigate metal as a conductor of heat.** Why do saucepans and cooking utensils often have wooden or plastic handles? Feel water pipes, radiators, 'warm up' coins in your hand. Put a metal spoon in a container of hot water and feel the end of it.

Metals and Magnets

- Investigate metals as conductors of electricity (see BATTERIES AND BULBS).
- When do we need to bend and shape metals? How is this done, e.g. making horseshoes, wrought iron, bending pipes, silver 'hammering', beating dents out of a car, making springs?

Creative Activities:

- Make close observational drawings of metal objects, e.g. brass musical instruments, kettles, gates and railings, inside a clock, machines and tools, metal objects in the kitchen.
- Take rubbings from gratings, grills, drain covers, radiators, coins, decorated metal work, brasses, scissors, keys, paper clips, bits of old clocks, metal buttons.

- Design and make badges.
- Design coins, jewellery, wrought iron patterns.
- Make 'junk' models using foil packaging.
- Make patterns in foil with a blunt pencil, or by pressing coins under foil.

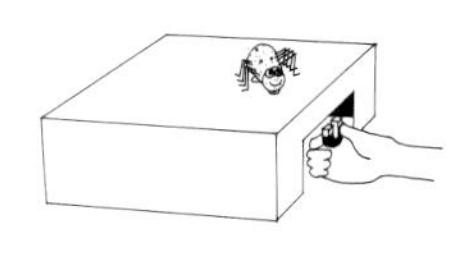

- Make a giant collage of 'The Iron Man' using boxes and tubes painted silver or covered with foil.
- **Make a magnetic theatre** (see photograph above). Use a shoe box or larger boot box for the base. Cut out end sections so that the puppeteers' hands can move the figures by moving the magnets underneath. The figures are made from corks, cotton balls or match boxes with drawing pins stuck into their base (paper clips or small magnets could be taped to the figures for very small children). Light the theatre with torches covered with coloured filters of Cellophane.

- **Make a magnetic mouse maze:** make a mouse out of a section of cork. Put drawing pins on the base of the mouse. Draw a maze on the base of a shoe box. Take the mouse to the cheese by moving the magnet under the box.

Batteries and Bulbs

Starting Points/ Discussion: Electrical things, batteries and bulbs as a topic will have been introduced in most cases as a development of others, e.g. from a consideration of materials such as wood, metals and plastics as conductors; from making toys that move; from lighting toys and models; from an investigation into mixing coloured lights.

The topic of electrical things provides a good opportunity for safety education. **Stress that for your investigations you will be using low voltage, safe batteries.**

⚠ **Never use mains electricity. It can kill.**

Never play with old batteries. Leaking chemicals can be dangerous.

Talk about the uses of electricity at home and at work. Electricity gives us light, heat, sound, pictures. It moves things and powers computers and machines. Talk about static electricity, and lightning.

Vocabulary: Power, energy, static electricity, lightning, current, volts, light, machines, electronics, filament, bulb, wire, motor, batteries, terminals, positive, negative, switch, circuit, conductor, insulator, electromagnetism.

Collections: Battery-powered toys, torches, radio, tape recorder, pocket calculator, games, bell; materials for making simple circuits, including torch bulbs and bulb holders, wires, batteries, aluminium foil, a small screwdriver, paper clips, paper fasteners, a small motor, and a buzzer. (Make sure that the voltage of the torch bulb matches — as closely as possible — the voltage of the battery, e.g. use a 1.25 volt bulb with a 1.5 volt battery, a 2.5 volt bulb with a 3 volt battery, or a 3.5 volt bulb with a 4.5 volt battery.)

Pictures of machines, night-time cityscapes, fairground and advertising illuminations, lightning flashes, traffic lights, etc.

Science Activities: Set a series of challenges and provide suitable materials.

- **Can you make the bulb light up?** Make a simple circuit making sure that all the connections are good or the bulb will not light. Introduce the vocabulary 'complete the circuit' and 'break the circuit'.
- Which materials will electricity pass through? Using a simple circuit and everyday objects, discover which materials are conductors and which are insulators.
- Look carefully at the disconnected torch bulb. Can you see what carries the electricity and also makes a light? Observe how thin and delicate the filament is. What do you think happens when a bulb 'goes'?
- Can you make a simple switch? Make a flashing light for a model. Supply materials such as foil, drawing pins, paper clips and fasteners, matchboxes, thick card or soft wood.
- Can you make a battery light two bulbs? Three bulbs? Can you find different ways of doing this? What happens to the lights?
- Look at the inside of a torch. Can you see how the circuit is made? Can you see how the switch works? Can you change the batteries? Does it matter how they are replaced? Look carefully at the terminals of the batteries. Can you make a pretend torch?
- Use torches and coloured Cellophane to make and mix coloured lights.
- Charge a balloon with static electricity (see IT'S PARTY TIME).

Creative Activities:

- Make maps of circuits made with batteries, wires, bulbs, motors and switches.
- Make 3D collages of simple circuits. Let children choose their own materials to depict wires, batteries and bulbs. (Supply Cellophane, clear plastic packaging, wools, twine and strings, boxes, tubes, foils.)

Batteries and Bulbs

- Make 3D models and toys. Use bulbs and batteries to light vehicles, a lighthouse, a toy theatre, a dolls' house, traffic lights, dragon or monster's eyes. Make a 'steady hand' game with a buzzer. Make a toy vehicle or model move with a small electric motor.
- Cut out pictures of electrical appliances and machines to make collages. Sort out according to use, e.g. light, heat, sound, moving things, and make a chart.
- **Make a montage of paintings or detailed drawings of appliances and machines found in the home.** Group them according to where they are used, and mount them overlapping to build up a house shape.

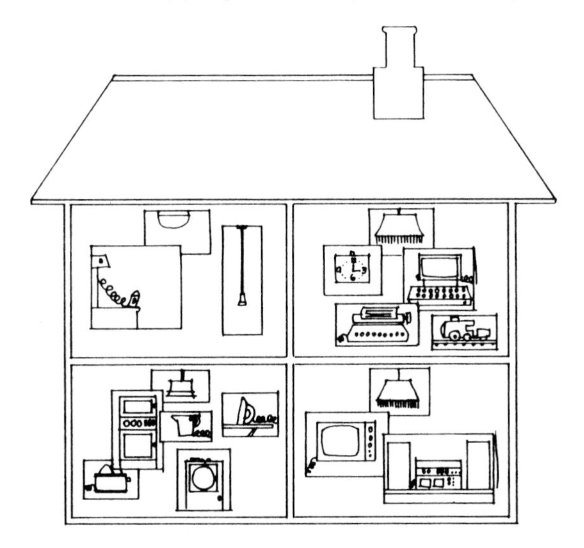

- 'Lights in the gloom' pictures. Use the wax resist or wax and scratch technique to make pictures of Christmas tree lights, lightning flashes, lampposts, lit windows of a cityscape, fairground lights, a lighthouse, etc.
- Lightning paintings — make zig-zag patterns with bright colours on dark paper. Mount on a foil-covered display board (see photograph above).
- Make a coloured lantern.

Floating and Sinking

Starting Points/ Discussion: The topic of floating and sinking could develop from a study of wooden things, plastics, etc. or from water-play activities. It could also develop from a story starter such as 'In which Piglet is entirely surrounded by water', by A.A. Milne. A series of challenges could be set such as 'Can you invent a way of sending a message in a bottle?', 'Can you devise a boat, from everyday objects, to rescue Piglet?'. The stories *Who Sank the Boat?* by Pamela Allen, Hamish Hamilton, *Mr Gumpy's Outing* by John Burningham, Picture Puffin, or *One-Eyed Jake* by Pat Hutchins, Picture Puffin, are all concerned with floating and sinking, while *Mr Archimedes' Bath* by Pamela Allen, Bodley Head, specifically introduces the idea of displacement.

Talk about learning how to swim. Some of the children will be using buoyancy aids in swimming lessons. Advanced swimmers and life-savers practise swimming wearing heavy, wet clothes. Why is this necessary?

Vocabulary: Float, sink, bob, submerge, buoy, buoyant, light, heavy, bubbles, afloat, waterborne, dive, sediment, plunge, drown, raft, pontoon, submarine, anchor, displacement, surface, deep, shallow, water-line.

Collections: Model boats, bath toys, swimming aids, floats for fishing; different materials to test, e.g. marbles, wooden beads, plasticine, rubber, plastics, metal objects, bottles and containers with stoppers, corks, etc.

Make an activity table — cover with a waterproof tablecloth; provide bowls or trays on which to stand wet objects. Cover all labels and notices with plastic. Use a see-through water trough if possible.

Science Activities:

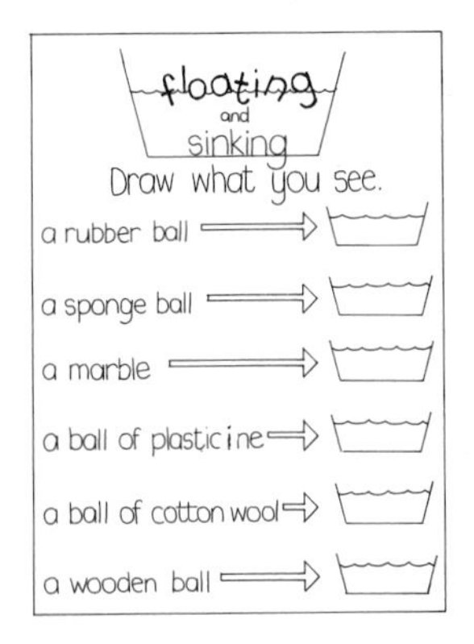

- Put everyday objects into the water. Sort them according to criteria — floats/doesn't float. Observe how the floater floats. How much can be seen above the water-line? Is it level in the water? Draw the position of the object in the water on a prepared sheet. Observe how the sinkers sink. Do they plunge straight down or do they go from side to side? Can you describe the pathway of your sinker through the water?
- Push down hard on a very good floater. What happens? Can you feel something pushing against your hand? What happens when you let go?
- Collect objects of the same shape but made of different materials, e.g. trays or lids (wood, metal, cardboard, plastic, polystyrene); spheres (a wooden bead, a marble, a ball-bearing, a ball of foil, a rubber ball, a sponge ball, a ball of plasticine, a cotton-wool ball). **Observe the position of the object in the water and record by drawing** (see worksheet example).
- Collect objects made of the same materials, e.g. wood, but of different shapes. Record what happens when you put them in water. Try plasticine or foil. Make different shapes (e.g. compare a *ball* of foil with a *sheet* of foil).
- Weigh objects before and after floating and sinking investigations.
- Observe floating objects over a period of time. Do you notice any difference in their positions in the water?
- **Make boats** with foil, balsa wood, paper, plasticine, polystyrene trays. Set a challenge. Who can make a boat that will carry six cubes, coins, play people? Read the story *Who sank the boat?* by Pamela Allen. Can you sink your boat? Devise different ways of propelling your boat through water.
- Try to float materials like cotton-wool, pieces of fabric, foil, polythene, feathers, bath sponges, grease-proof paper, blotting paper, paper towel, chamois leather, loofah, pumice stone, cardboard. Do you see any changes taking place? Push material to the bottom of the trough. What happens when you let go?
- Observe ice-cubes in water. How much of the ice-cube is under the water?

Floating and Sinking

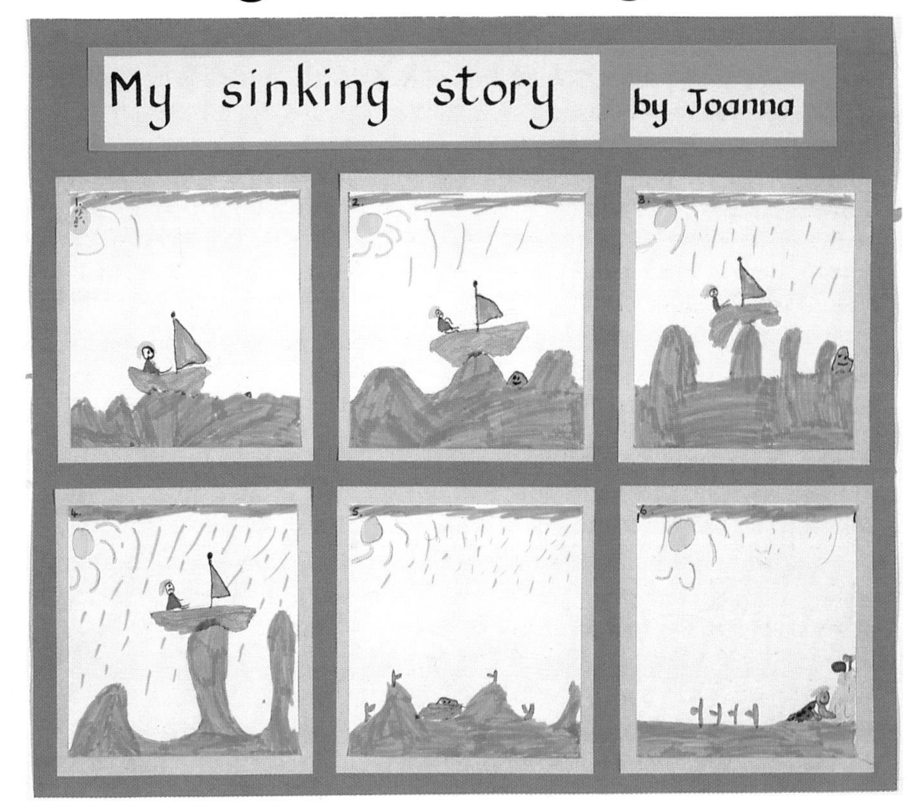

- Investigate displacement. Put your hand in a container full of water. Observe the water level. Put objects in a jar of water. What happens? How can you measure, or record, the changing water levels?
- Test floaters in different strengths of salt solution.
- Can you make a sinker float? How does a submarine work? Empty a soft drinks can or plastic bottle, fill it with water and allow it to sink. Can you devise a method of re-floating it? (Try using a straw or a length of plastic tubing.)
- Try floating plastic containers that have holes in them. Do they sink? Press down. Do all things with holes sink?

Creative Activities:

- Make a cartoon strip to illustrate a sinking story (see photograph above).
- Look at decorations and patterns on canal barges.
- Design and make a bath toy that floats. Consider materials to be used — glues and paints need to be waterproof. Toys should have no sharp edges or splinters.
- Make a 'bottom of the sea' collage of coral, creatures, shipwreck, anchors, divers, etc. Stipple brush the water — darken the shades of paint as the water gets deeper.
- Ideas for backing paper for 'water' pictures:
  - roller printing: using different shades of blue and green paint make a wave pattern with a roller
  - paint stripes of blue/green shades on wet paper
  - tear or cut strips of blue tissue paper and glue to backing paper. Print or paint on top of tissue strips
  - bubble prints (see front cover)
  - make a cardboard comb and draw it through layers of thick paint to make wave patterns
  - cut strips of blue shades from magazines to make collage background.

Building

Starting Points/ Discussion: A good point to start with would be the story of *The Three Pigs.* What materials did they use to build their houses? Which house was the strongest? Why? Which materials are used to build houses? Go for a walk around the school and look for all the different materials used — brick, stone, wood, glass, plastic etc. How are bricks put together? What patterns do they make, and how are corners formed? What is used to keep the bricks together? Invite a builder into the school or arrange to visit a site. What tools does a builder use? How does he carry bricks? How does he know what to build and where? Look at spirit levels and plumb lines — how are they used? Look at real buildings or pictures of different kind of buildings — office blocks, blocks of flats, sky-scrapers, churches, stately homes, warehouses, schools, factories, hospitals etc. Look at the different materials used. Which are made from concrete, brick, stone? Look at the shapes of roofs and pictures of houses from other countries. Discuss the influence that climate may have on the design of buildings. Look at the different shapes used in buildings — squares, rectangles, triangles. Read *Miss Brick the Builders' Baby* Allan Ahlberg and Colin McNaughton, Puffin Books/Viking Kestrel.

⚠ **Talk about why building sites are potentially very dangerous.**

Vocabulary: Stability, structure, foundation, concrete, bricks, breeze blocks, girders, scaffolding, pulleys, cranes, right angle, gutter, drainpipe, sewer, drains, insulation, architect, plumber, electrician, carpenter, bricklayer.

Collections: Builders' tools: plumb line, spirit level, trowel, pulleys, pictures of building sites, cranes, wheel barrows etc.; building materials: bricks, stone, cement, sand, chimney pots and piping (contact a local builder who may be happy to help out with some of these items); pictures of different kinds of buildings; construction kits e.g. Lego.

Science Activities:

- Build a variety of walls from junk materials or building blocks. How strong is your wall? Design a fair test to find out. (Run a toy car down a slope into the wall, or swing a ball of plasticine on a piece of string.)
- Investigate the need for foundations. Try building on different surfaces, e.g. in the sand tray, using wet/dry sand.
- Can you make some bricks? Using a variety of natural materials, e.g. clay, sand, soil, stones, straw etc., find out which makes the best bricks. Mix any of the soils with the stones and/or straw. Make a mould — use a large matchbox or make a wooden mould. Decide on how long to leave them to set. Decide upon a fair test — e.g. will they break if dropped from a certain height? How many weights (standard or non-standard) can you place on them before they crack?
- How do builders lift heavy objects? Look at cranes, pulleys etc. Can you make a pulley that will lift a wooden building block from the floor to desk height?
- Develop the previous idea further — can you make a crane which will move an object from one place to another — up and down *and* side to side.
- **Do bricks absorb water?** Obtain some bricks and stand them in a tray of water. Observe what happens. How can you stop water soaking up a wall? Make a damp course — investigate different materials.

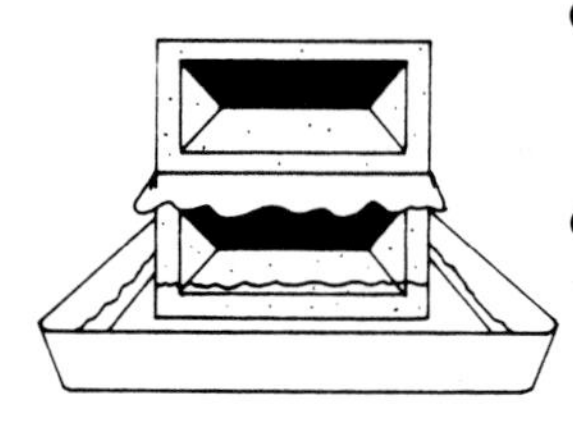

- Look at roof shapes. Design and make model houses with roofs suited to different climates — heavy rain, snow, wind, a lot of sunshine. How can you make a roof waterproof? Investigate using different materials. How can you simulate rain (droppers, watering can etc.)?
- Test the strength of different shapes, e.g. squares, triangles, rectangles. Using straws or geostrips, design a fair test to find out which shape is strongest.
- Build with drinking straws or art straws. Use pipe cleaners to secure ends. Tubes are often used in building. Investigate the strength of different tubes — newspaper, straws etc. (see BRIDGES).

Building

- Provide a light source in a model of a building, (see BATTERIES AND BULBS) e.g. shoebox or dolls' house.

Creative Activities:

- Make junk models of different kinds of buildings — sky-scrapers, office blocks, houses, high-rise flats, shops, hospitals etc.
- Take rubbings of different building materials and use these to make collage pictures of buildings.
- **Card rubbings — cut out pieces of card and make pictures of buildings.** Glue on to another piece of card, overlaying the details. When stuck securely, place a piece of paper over the top (kitchen paper is best), ask someone to hold it, and rub over with wax crayons. Display with card blocks (see photograph). A wash can be put on over the top to provide a contrast.
- Skyline pictures. Cut out roof shapes from a piece of black paper. Glue on to another piece of paper which has been washed over with sunset colours.
- Make a street collage of different buildings. Use boxes to give a 3D effect. Put a light source in each box so that the lights can be put on to simulate night time. Windows will look effective if covered with thin material, or Cellophane or tissue paper.
- Printed buildings — use sponges cut into rectangles for the bricks. Use other blocks for roof tiles, chimneys etc.
- **Build sandcastles and decorate with flags, lollysticks, matchsticks, shells and stones.**

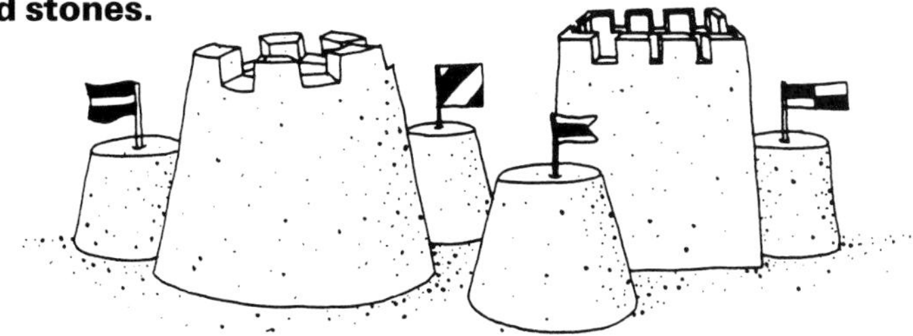

On a Windy Day

Starting Points/ Discussion: Ask 'Can you *see* the wind'? This leads to a consideration of the effects of the wind. Go outside and feel the force of the wind on your face and hands. Turn around and feel the difference. Can you find a sheltered place? Talk about the evidence of the wind blowing, e.g. effect on a weather vane, hair, washing on a line, litter, long grass, trees, pets' fur, clouds, smoke, dust, raindrops, puddles or pond. Can you find the direction of the wind by looking at these things? Use streamers, flags, a wet finger to test. Try and 'catch' the wind in a carrier bag or a pillow case. Blow bubbles, or float feathers and see how far they travel before hitting the ground. Try and talk to your friend from some distance. Run against or with the wind.

Read *The Wind Blew* by Pat Hutchins, Picture Puffin.

Vocabulary: Gust, breeze, hurricane, whirlwind, gale, air currents, airstreams, draught, squall, waft, howl, flurry, tornado, cyclone, windmill, sails, windproof, Beaufort Scale, anemometer, weather vane, wind-sock, fresh, strong, light, moderate, calm, gale force.

Collections: Toy windmills, balloons, shuttlecocks, kites, decorated fans, musical instruments (brass and woodwind), mobiles, model gliders, seeds, pictures of sailing ships, parachutes, gliders, windmills.

Science Activities:

- Can you discover the direction of the wind? Make a simple wind-sock.
- Can you find a way of measuring the strength of the wind?
  - find out about the Beaufort wind scale and the signs used to denote wind force
  - observe different materials blowing in the wind e.g. different weights of fabric and paper suspended by thread from a dowel. Ask the children to suggest materials. Look at the extent of their movement.
  - **investigate making a calibrated instrument to measure the force of the wind.**

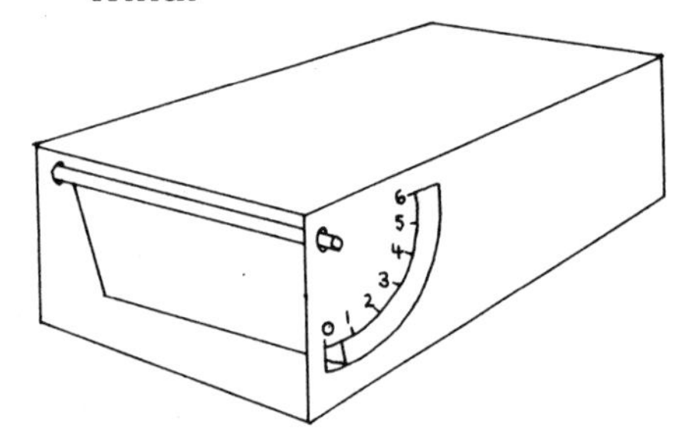

- Observe and record weather conditions over a period of time. Are there more wet and windy days than dry and windy days? Record temperatures and look for patterns.
- Find windy places, sheltered places. Can you see any pattern? Try different heights of ground. Can you design an effective wind-break? Why do people often use wind-breaks on the beach?
- How can we make a wind indoors? Try flapping, fanning, blowing, a balloon pump, squeezy bottle. Try to move things by blowing them, e.g. different papers, feathers, etc. Devise fair tests to find out the effects of wind: drying fabrics, cooling things.
- ⚠ Observe the movement of things through the air. **Drop things and watch how they fall,** e.g. feathers, corks, plasticine shapes, paper, balloons, parachutes, shuttlecocks.
- Investigate dispersal by the wind (e.g. dandelion clocks). *Charlotte's Web* by E.B. White, Puffin, contains a graphic description of the dispersal of spiders by the breeze.
- Investigate how wind power is harnessed and used, e.g. windmills, sails, wind-surfing, kites and gliders. Make paper windmills, gliders and kites.

On a Windy Day

from *Flat Stanley* by Jeff Brown, Magnet.

- Find out about wind-proofing. Which are the best fabrics?
- Find out about musical instruments that you blow — woodwind and wind instruments.
- Investigate smells and scents carried by the wind. *Fantastic Mr Fox* by Roald Dahl contains a good description of how the fox locates the farmers by using the wind.

Creative Activities:

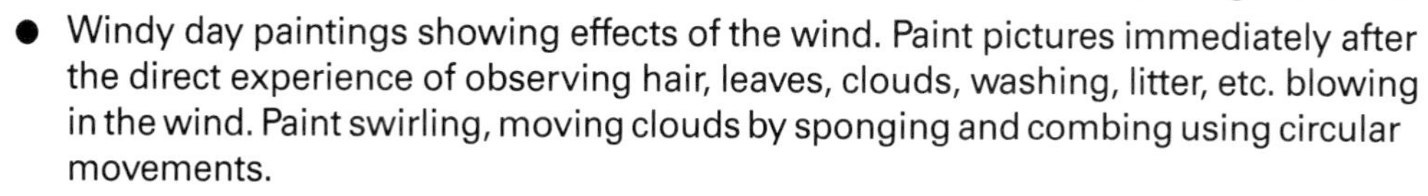

- Windy day paintings showing effects of the wind. Paint pictures immediately after the direct experience of observing hair, leaves, clouds, washing, litter, etc. blowing in the wind. Paint swirling, moving clouds by sponging and combing using circular movements.

- Illustrate a 'windy day' story, e.g. Flat Stanley being flown as a kite (see photograph above).

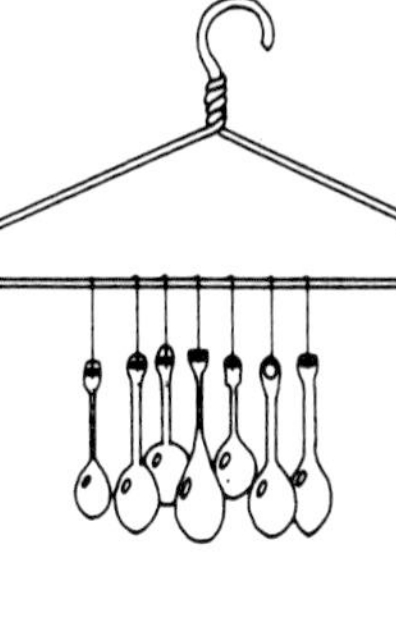

- Design flags. Use paint and crayon with paper, and fabric crayons to decorate fabrics. Consider how the flag is going to be fixed to the flagpole — design a pulley system to fly your flag.
- Make and decorate a kite.
- Make and decorate a fan. Look at a collection of fans from different countries. All kinds of decorations can be used, e.g. folding, cutting, water colour, embroidery, etc. according to materials used.
- **Make a mobile that will move gently in a breeze.** Can you make a wind chime?
- 'Blown' pictures. Using straws make a blown ink picture or make blown, dry powder paint pictures — keep turning the paper.
- Design a litter bin for a windy seaside town.
- Make a decorated paper windmill.

Ice and Snow

Starting Points/ Discussion: Wait for a snowy day! Go outside and play in the snow. Talk about suitable clothing, the need for waterproof footwear, gloves, hats, warm coats. Make footprints and tracks in the snow. Listen to the sound of feet compacting snow as you walk. Look for animal/ bird tracks in the snow. Look at the changed landscape — the changed shape of familiar things. Can you find a snowdrift? Why is the snow deeper in some places than in others? Can you find a way of measuring the depth of snow in different places? Can you tell which way the wind was blowing as the snow fell?

Find some ice. Look in hollows, on window ledges, the pond. Compare with snow. Can you break it? Listen. Look at shapes of icicles. Make snowballs, snowmen, etc. What happens to your gloves? Make a ball head by rolling a snowball. Can you make different shapes? Snow pies? Take advantage of any chance to observe snowflakes, hail or sleet falling. Look at the pathway of the snow as it falls — is it straight down or slanted? What happens when it hits the ground? Can you hear anything? Does it bounce? Does it melt or 'settle'? Consider the slipperiness of ice, the best shoes or boots for walking on ice, and the gritting of paths and roads. Come inside. What happens to the snow from your shoes/boots? What do you notice about your fingers? Are they 'tingling'?

Read *The Snowy Day* by Ezra Jack Keats, Picture Puffin; *The Snowman* by Raymond Briggs, Hamish Hamilton.

Vocabulary: Solid, liquid, freeze, melt, thaw, icicle, iceberg, crystal, frost, snowdrift, sheets, bleak, wintry, stark, chill, sleet, hail, slush, flurry, blizzard, Arctic, glacier, shivering, chattering, polar, avalanche, ski, skate, sledge, toboggan, slide.

Collections: Pictures of wintry conditions, mountains, winter sports, glaciers, polar animals; warm winter clothing, e.g. ski gloves, hats, scarves: sports equipment, e.g. skates, skis, snow-goggles, a sledge, ski-boots.

Science Activities:

- What happens if you put snow in water? Observe water level in container.
- Collect a container of snow and weigh it (using non-standard or standard units). Leave snow to melt and then weigh it again. Is there a difference? Observe water level in container. Make a chart to record results.
- Investigate how you can stop snow/ice from melting inside. Refer to *The Snowman* by Raymond Briggs. Investigate 'wrapping-up' and insulation.
- Investigate where ice and snow melt most quickly. Leave in different parts of the room/building. Devise a fair test — consider size or amount of snow and ice, container/surface.
- What happens if you put salt on ice? Discuss clearing roads of snow.
- **Freezing. Make ice cubes and ice lollies** (how can you make different flavours?) What do you notice about the level of ice in the container (fill to the top before freezing)? Leads to discussion about frozen pipes cracking and floods when there is a thaw.

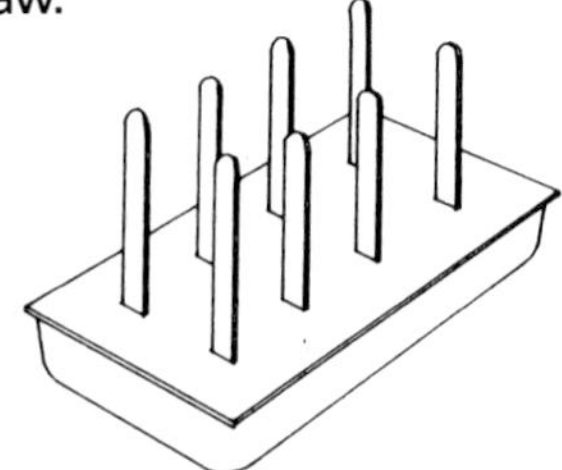

- Put ice cubes in water. What can you see? *Where* are they floating in the water? Discuss the dangers of icebergs to shipping. Try and push your ice cubes to the bottom.
- Ice is slippery. Slide ice cubes across different surfaces. Look at the treads of boots and shoes. Which are best for walking on ice and snow?

Ice and Snow

Creative Activities:

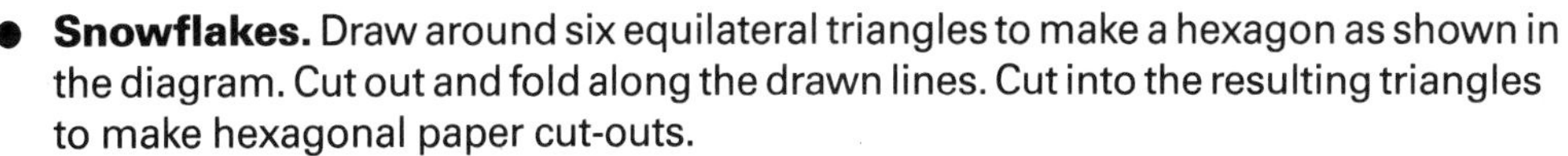

- **Snowflakes.** Draw around six equilateral triangles to make a hexagon as shown in the diagram. Cut out and fold along the drawn lines. Cut into the resulting triangles to make hexagonal paper cut-outs.

- Make card printed snowflakes for backing paper or borders (see photograph above).

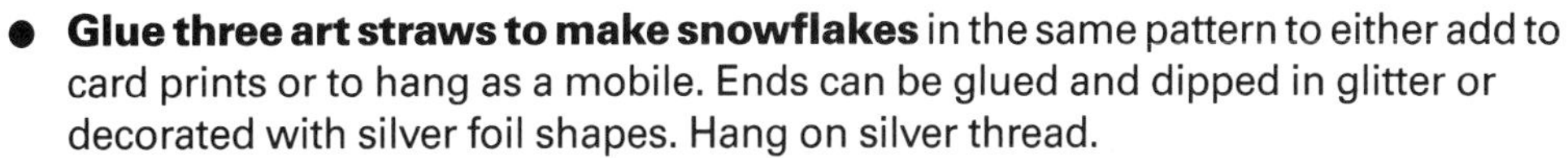

- **Glue three art straws to make snowflakes** in the same pattern to either add to card prints or to hang as a mobile. Ends can be glued and dipped in glitter or decorated with silver foil shapes. Hang on silver thread.

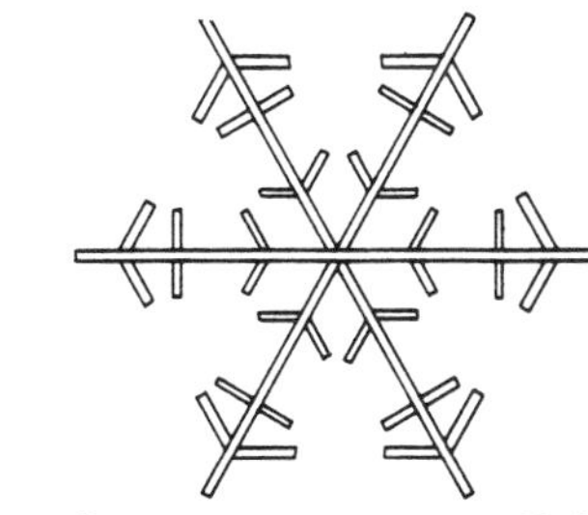

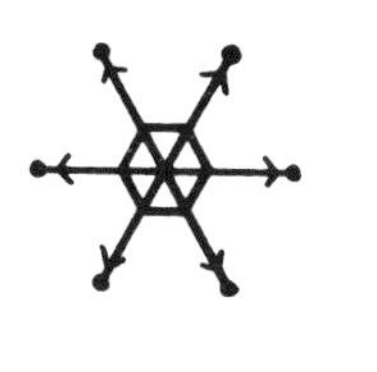

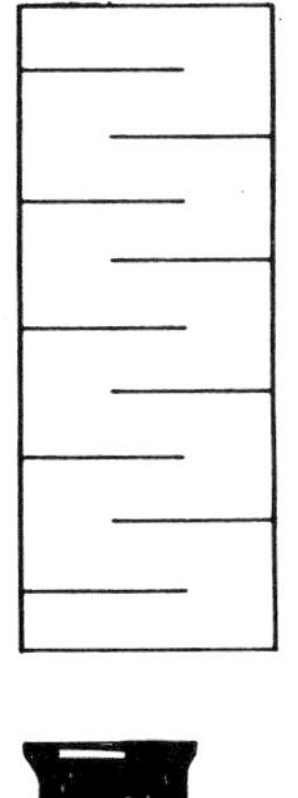

- Snowman cut-outs. Fold paper in half and draw half snowman on fold (young children need practice at this). Cut out in one piece. Cut-out snowman can be used for masking and sponging. Children can make their own repeated frieze. Outer stencil can also be used as a tissue paper window frame — use white, pale blue, purple overlapping tissue paper circles.

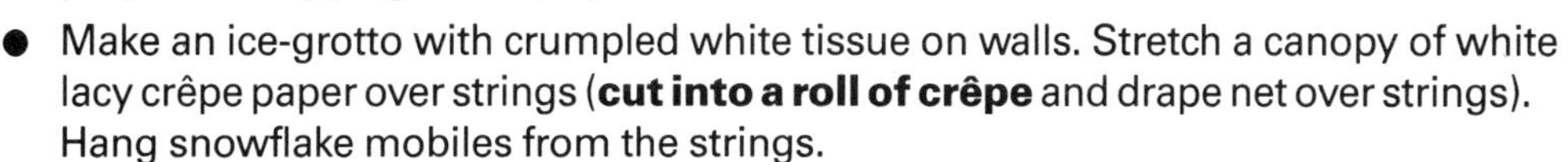

- Make an ice-grotto with crumpled white tissue on walls. Stretch a canopy of white lacy crêpe paper over strings (**cut into a roll of crêpe** and drape net over strings). Hang snowflake mobiles from the strings.
- Fix doily to backing paper with masking tape. Sponge white paint carefully to make prints.
- Sponge or roller a snowy background — ice blue and white. Tear or cut black shapes to create silhouettes. Glue onto backing.
- **Make a snowman book.** Record the date and time of making the snowman. Write daily observations. Draw pictures of stages of melting.
- Make 'tracks in the snow' pictures by printing with the treads of boots and shoes.

Puddles and Rain

Starting Points/ Discussion: What happens when it rains? Rain falls in different ways — small drops, drizzle, heavy drops, hail. Sometimes the wind blows the rain and we hear it hitting the windows. Where else can you hear rain (roofs of cars, roofs of houses, on caravans)? Where does the rain come from? What can't we do when it's raining (play outside, have a picnic, etc.)? How do we know when it has been raining? What happens if there is more rain than normal (floods, rivers bursting banks, etc.)? Why do we need rain? What else needs rain? Talk about deserts and jungles. Look at gutters, roofs, drains. Discuss what happens if things are left out in the rain (toys, bicycles, clothes on the line).

Vocabulary: Absorbent, non-absorbent, evaporate, rainbow, water cycle, waterproof, flood, splash, droplet, soak, saturated. Sayings about rain — 'raining cats and dogs', 'bucketing with rain', 'raining stair rods'.

Collections: Umbrellas (take care with spokes); Wellington boots; waterproof hats, coats, trousers; pieces of guttering, water butts; seaweed (for rainy day detecting); rain gauges, weather forecasts and symbols for rain; stories/poems about rain.

Science Activities:

- Can we make puddles? Make puddles of water on different surfaces, e.g. soil, playground, sand, grass, polished surface (lino — but not in the corridor!). What happens to the water? What does this tell you about the surface?
- What happens to puddles that do not soak into the ground? After it has rained, go looking for puddles. When a suitable sized one has been found, draw around it using chalk. Depending on weather conditions, leave a suitable time then return to the puddle. Has it changed at all? How can you tell?
- Drying clothes: why do we hang wet washing out to dry? Does the water all drip off or does something else happen? Investigate with pieces of wet cloth in different locations.
- Getting wet: what happens if we leave things out in the rain? Investigate the effects of water on metal, crêpe paper, newspaper, bread, cardboard, etc.
- Soaking up puddles: investigate which materials soak up water the best. Make the puddles the same size each time — measure the water. Try paper towels, kitchen sponges, towels, etc.

- **Painting with water:** using a bucket of water and a large brush (outside), make some 'water paintings' on walls or playground. What happens?
- Spreading out puddles: how can we speed up the drying of a puddle? (Puddles can be inconvenient in playgrounds, etc.) Investigate the effects of brushing out the puddle to make it wider and shallower.
- Using a hand lens observe water drops on different surfaces, e.g. newspaper, polished desk, carpet, finger, etc.
- How can we move a model boat across a puddle? Design a sailing boat. How will you keep the sail up?
- Will plants grow without water? Try growing seeds with and without water.

Creative Activities:

- Paint pictures on wet paper and compare with pictures painted on dry paper (see photograph).
- Ink drops. Soak good quality paper, drop coloured inks onto the paper — watch the inks spread.
- Straw blowing. Make puddles of ink on paper and blow into interesting shapes; make a forest or a spiky creature.
- Marbling. Use marbling inks or mix powder paint with vegetable oil.
- Mono prints. Make a puddle of paint on a shiny surface. Make patterns in it using fingers, combs, etc. Place paper on top to make a print.

Puddles and Rain

- Rainy day mobiles. Use card to cut out shapes. Stick strings to top. Balancing may be difficult. Hang them on to coat hangers or wooden rods attached in a cross shape.
- **Multi-media collage of rainy day scene.** Sponge background — lots of grey for clouds. Paint in rainbows, etc. Cut out umbrellas from bright fabrics and stick on. Cut out clothes from fabric. Cut out puddles from Cellophane.

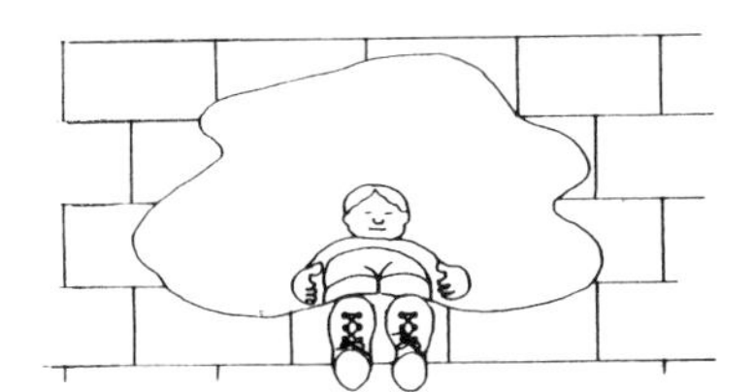

- **Reflections in puddles.** Cut a puddle shape from foil. Paint your reflection, cut out, glue onto puddle.

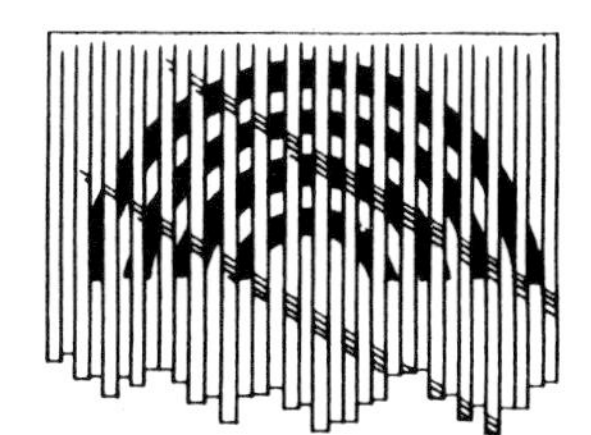

- **Raindrops hitting a puddle.** Paint concentric circles — blues, whites, greens and greys.
- **Rainbows through the rain.** Cut silver foil into strips, with cut-out rainbow interwoven through 'rain'.

Healthy Living

Starting Points/ Discussion: Begin by asking the children what they think is necessary for them to have and to do if they are to grow up healthy and strong. List their suggestions, e.g. taking exercise, keeping clean, eating, drinking, going to bed early, love and care, keeping away from germs. Sometimes we need help in keeping healthy. Discuss visiting the dentist, optician, doctor, school nurse, etc. Discuss the illnesses that the children may have had. How did the doctor help them get better? This is discussed further in 'Keeping Safe'. Invite the school nurse in to talk about keeping healthy. Read *The Toothbrush Monster* by Rose Impey and Sue Porter, Picture Puffin. How do we feel when we are healthy? Happy, strong, fit, full of energy. How do we feel when we are unhealthy?

Vocabulary: Health, hygiene, shampoo, soap, germs, dirty, unhealthy, exercise, running, skipping, swimming, jumping, washing, brushing teeth, heart, lungs, blood, pulse, muscles, sweating, dangerous, safe, hazard, doctor, nurse, dentist, optician, medicine, injection, pills, treatment, prescription, fruit, fibre, vegetables, wholemeal bread, foods for energy, foods for growing, vitamins, minerals.

Collections: Contact local Health Education Unit for health related resources. Posters, stickers and free samples from dentists; bars of soap, face flannels, towels; pictures of people taking exercise — running, swimming, playing football, skateboarding, etc; a stethoscope; pictures or packets of 'healthy' foods; pictures of doctors, nurses, etc. (Contact ROSPA for further information on keeping safe).

Science Activities:

Hygiene and Fitness

- What happens when we take exercise? Count breaths while resting, feel the pulse beating in your neck. Get the children to decide on some exercises e.g. skipping, running, jumping, dancing, etc. After a suitable period of exercise count breaths again, feel heart beat and feel pulse beating in neck. Have the children started to sweat? Do they feel hotter? Make simple before/after block graphs, illustrating how each child's breathing rate changes with exercise.
- What does sweating do? Why do we sweat when we get hot (to cool us)? Place one hand in water and then wave both hands in the air. Observe what happens to the water on the wet hand. How does the wet hand feel?.
- We need to avoid germs to keep healthy. Where do germs live? Germs love warm, dirty places. Where might germs live on us? How clean are our hands? Look very carefully at nails, under nails, creases in hands, etc. with a hand lens. How can we get our hands clean? Investigate the effects of washing dirty hands with cold water, hot water, cold water and soap, hot water and soap, etc. Which method was most effective?

Eating and Drinking

- We all need to eat and drink to stay alive. Collect packets and pictures of foods. Sort into broad groups, e.g. main part of meal, fruit and vegetables, bread or cereal. Discuss favourite foods, avoiding the terms 'good' or 'bad' food. Plan meals for a day to include: foods for growing, foods for energy, foods for vitamins, etc. Suggest some healthy alternatives to sweets and snacks, e.g. low fat crisps, carob bars, cereal bars, fruit, vegetables and sprouted seeds.

Keeping Safe

- Using a plan of a room in a house, discuss with the children all the areas that are potentially dangerous to a young child. (The kitchen would be a good room to use.) Talk about pot handles, ovens, irons, medicines, cleaning bottles, dirty places, electric sockets, etc.

Healthy Living

- Discuss the safe use of medicines. Who gives them medicine (doctor, nurse, mother or father)? Explain that medicines only make you better if taken properly — when needed and in the correct amounts — but very ill if taken wrongly. Only adults can give them out.
- How much rest do we need? Young, growing creatures need more rest than adults. Make a bar chart to show the different amounts of sleep that people need at different ages.

Creative Activities:

- Movement pictures. Make jointed pictures with card and paper fasteners. Arrange figures so that they are playing different sports (see photograph above). These could be used as stencils for sponging or spraying.
- **Bedtime pictures.** Make quilts using cut-out hexagons from wallpaper or paper (cutting fabric can be difficult) or make paper-weaving blankets. Cut out head and hands and put at the top. Use wool for hair.
- Make a model of a fantasy germ. Use clay or plasticine or junk materials. Make it look as ugly and gruesome as possible. Give your germ a suitably horrible name and say where it would be found.
- 'My Healthy Day'. Make a zigzag book — draw pictures of the healthy things done in a day.
- **Design a poster** to encourage people to wash their hands, clean their teeth, eat healthy snacks, etc.

Bones and Movement

Starting Points/ Discussion: What can you feel under your skin? Feel head, ribs, hips, knees, ankles and wrists. Why do we have bones? Ask a child to put on a coat then take it off — let the coat drop to the floor. Our bodies would be like the coat if we did not have our skeleton. Skeletons protect our delicate insides (ribs, skull). Feel where arms and legs bend. How many different parts of yourself can you bend or move? Which joints move like hinges — elbows, knees, fingers. Which joints swivel — wrists, ankles, necks. How do we move our bones? Feel the muscles in arms and legs. Make the muscles bunch up — 'look strong'. Muscles pull our bones so that we can move. Discuss which sportsmen use particular muscles — swimmers, runners, tennis-players, weight-lifters. Have you ever broken a bone? What happened when you went to hospital? What sort of bones do animals have? We sometimes find bones in the meat we are eating at mealtimes. Some animals have shells on the outside of their bodies (see MINIBEASTS). Bones and shells are made of similar material. Talk about fossils of animals that lived long ago.

Vocabulary: Light, dry, white, hard, hollow, lacy, sharp, rounded, ball and socket, swivel, hinge, muscle, backbone, spine, vertebrate, invertebrate, skull, skeleton, ribcage, smooth, shiny, slippery.

Collections: Secondary schools and museums often lend cleaned bone collections. Bird bones, skulls, fish bones, chicken bones must be cleaned, boiled and disinfected. Also useful are X-rays — hospitals will give these away; Halloween skeletons from joke shops — inflatable skeletons are available; pictures of food rich in calcium for growing bones; pictures of sportsmen/women showing movements and muscles; pictures of dinosaurs, etc. to display with fossil collections; animal X-rays or pictures of animal skeletons — a vet may help with these.

Science Activities:

- Making joints. Start with strips of thick card. How can the pieces be joined together so that they will bend? Use adhesive tape, thread or paper fasteners. Try using Meccano, geostrips or similar, to investigate joints.
- **On a picture label all the bones/parts you know.**

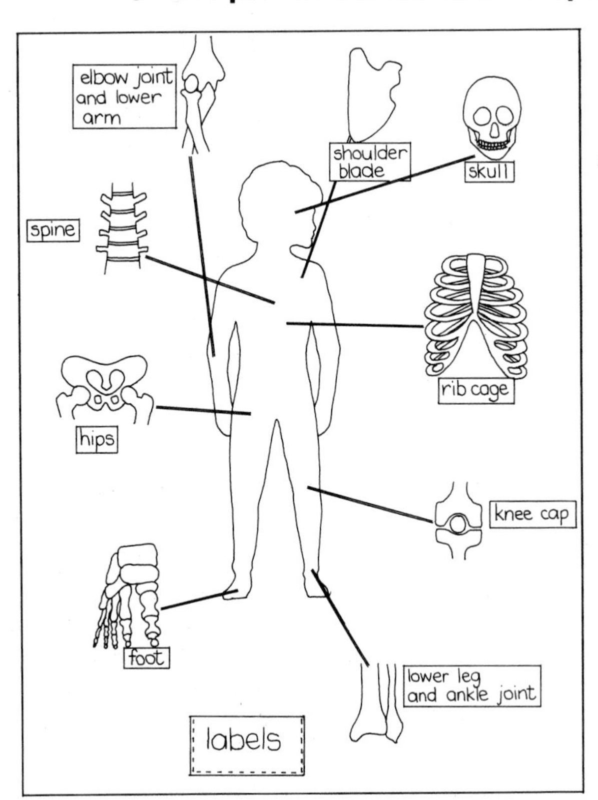

- Why are bones strong? Most long bones are cylindrical. Are cylinders strong? How much weight can different card cylinders take? Design a test. How would you place ⚠ the weights? Do this on the floor — **make sure fingers and toes are away from falling weights.**

Bones and Movement

- Animals with and without backbones. Make a model backbone using cotton reels, beads or pieces of art-straw threaded on a shoelace. Animals with backbones are called vertebrates. Animals without backbones are called invertebrates. Can you name some? Make a chart to show which animal fits into each category.

Creative Activities:

- Make an X-ray. Using tracing paper and white wax crayons, make a skeleton picture. Remember, only the bones must be drawn (look at real X-rays and real bones first). Wash over with thin black ink or paint. Display on the windows. Are the bones broken? Write the names of the patient, and the doctor and the date.
- Close observational drawings of bones — skulls are particularly good. This should give an opportunity for shading — don't leave the eye holes blank. Display on black paper, or do drawings with chalk and display on white. Display with bone collection on a black drape.
- Using art straws, polystyrene chips, pieces of paper, make a skeleton collage. Read *Funnybones* by Janet and Allan Ahlberg, Picture Lions, for ideas, particularly the visit to the zoo. Use black paper for background (see photograph above).
- Make a fossil. Roll out plasticine to approx. 4cm in depth — start with a sphere. Press a bone or shell or real fossil into the plasticine — remove. Make a card 'skirt' around the plasticine, making sure it is a tight fit. Secure it. Make some plaster of Paris and pour into the mould. Leave to dry. Peel off the plasticine. Paint it. Experiment with bones, ferns. This could lead to creative writing — what is your fossil? Where did you find it? How long ago did the creature live?
- Make a skeleton which has joints that move. Use paper fasteners for the joints.

Ears and Sounds

Starting Points/ Discussion: Begin by making loud noises — bang a drum, drop a metal plate, turn a cassette player on with high volume. Then make soft noises — whisper, drop a pin, drop some paper. Discuss other loud noises, for example warning sounds, sirens, horns on cars, bells; sounds to tell us something — lunch bell, alarm clocks, alarms on watches, timers on cookers. How do we hear all these different sounds? Look at ears — what is the outside part for (collecting sounds)? What is the hole for? Where does it go? ⚠ **Stress safety point here about not putting things into ears.** Talk about the delicate eardrum. **Very loud noise can damage ears.** Try listening to someone talking while your ears are covered. Can you make your ears work better if you put your hands behind them and cup them? Bring some pets into the classroom or look at pictures of animals. Look carefully at their ears. Are ears all in the same place on all animals? Have all animals got ears? Which animals prick their ears up or turn their ears? Why do they do this? Listen to the 'sea' in shells.

Vocabulary: Loud, soft, vibrate, bang, crash, twang, megaphone, earphones, quiet, pitch, tone, rhythm, timbre, melody, music, decibel, clang, clatter, whoosh, fizz, buzz, crack.

Collections: Tape recordings of sounds — at home, at school, in the streets, etc.; telephones, alarm clocks, bicycle horns, tape recorders, radios, earphones, megaphones, personal stereos with earphones, bells, gongs, chime bars, drums, cymbals, guitars, shakers, rattles, bottles with different levels of water, rulers, elastic bands on boxes, spoons; ear muffs (protecting ears from loud noises); pictures of animals showing their ears; old radios or loud speakers showing amplifiers; hearing aids (or ask a nurse to come to talk about people who cope with hearing problems), records and a record player.

Science Activities:

- 'Seeing sounds'. Sounds are caused when the air is made to vibrate. When the vibrations reach our ears we hear the sound. Twang rulers on the edges of desks, see and hear the vibrations. Twang elastic bands over boxes. Place a 'charged' tuning fork in water — watch the water move. Hum a note and feel the vocal chords vibrate.
- What happens if we stop the vibrations? Place the ruler at different positions overhanging the desk. Why won't it make a noise when only a small amount is overhanging? Twang the elastic bands over the boxes while holding on to the bands with the other hand. Strike a triangle while holding on to the metal part — then strike it again while holding it correctly. Investigate further with guitars, drums, bells, etc.
- Hearing through different objects and materials. Can you hear a ticking clock through the desk? Can you hear footsteps with your ear pressed to the ground? Can you hear a ticking clock through a tank of water? Try striking water pipes in the classroom — how far away can the sounds be heard?
- Tie spoons to the ends of pieces of string. Jangle the spoons and put the strings near the ears. What can you hear?
- How far away can you hear different sounds? Choose a selection of sound makers — pin dropping on a table, coin dropping in a metal lid, or a cassette player. How can you ensure that the sound is exactly the same each time (drop object from the same height/use volume control on cassette player)? Mark out distances away from the sound source. Record which sounds can be heard if either ear is facing the sound source.
- **Make a sound collector** and repeat the above — does the sound collector work?

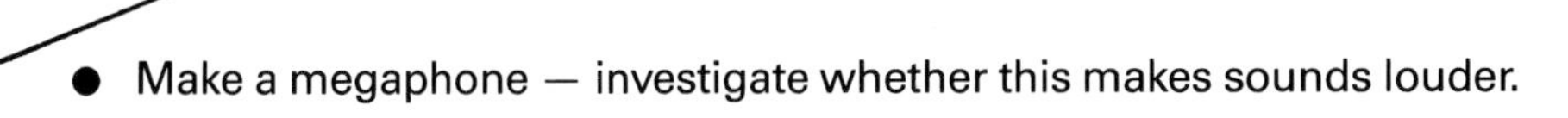

- Make a megaphone — investigate whether this makes sounds louder.

Ears and Sounds

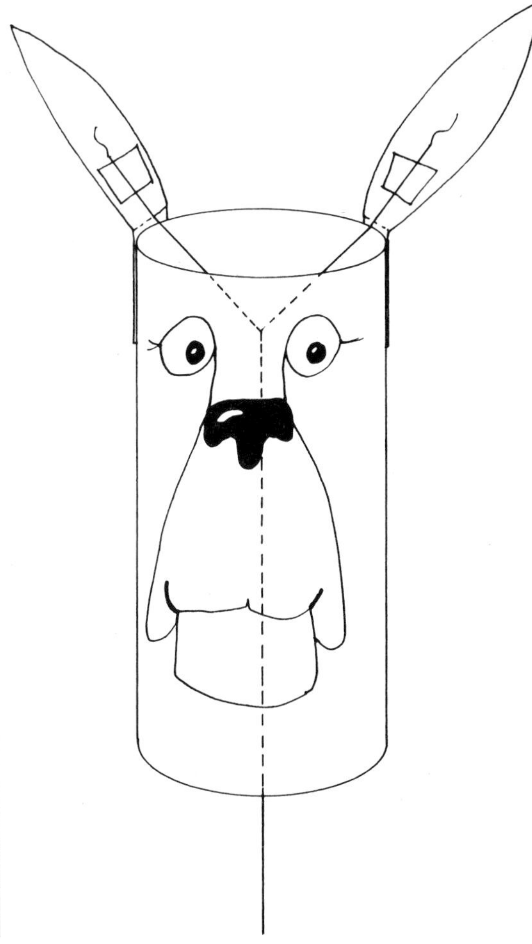

- Making sounds travel. Make a string telephone. Investigate which endpieces are best — yoghurt pots, tins, boxes (cuboid). Try using cardboard cylinders and covering the ends with different material (make sure it is tightly stretched over the end). Tie matchsticks to the ends of the pieces of string so that they don't get pulled through the holes. Investigate different lengths of string — does the string have to be taut? Try using funnels pushed into the ends of plastic tubing. Try whispering down the endpieces.
- **Make a stethoscope.** Either use the funnels as above, or cut off the ends of washing-up liquid bottles and push on to plastic tubing. Listen to the sound your heart makes.
- Stopping sounds. Can you hear as well with a thick woolly hat over your ears? Investigate.
- How can you soften the noise made by a loudly ticking clock, while still being able to read the time? (Wrap it in a scarf or similar material.)

Creative Activities:

- Choose a favourite story and make a tape of someone reading it. Add sound effects.
- **Make a model of a dog's head.** Attach strings which make the ears prick up.
- Observational drawings of musical instruments. Obtain some instruments from other countries to investigate and draw.
- Make musical instruments from junk materials. Paint and decorate. Have a recital using homemade instruments.
- Sound pictures. Graphical interpretation of sounds, e.g. twang, crash, ping, etc. (see photograph above).

In the Kitchen

Starting Points/ Discussion: Look at and discuss the contents of the different cupboards and storage places for food. Look at dried, tinned, preserved/pickled, chilled, frozen and fresh foods. Compare foods of the same kind that have been treated in different ways. Look at colour, texture, shape. Do they smell different? Discuss why some foods will keep longer if stored in different ways. Look at 'sell-by' dates and instructions for storage on packets. Make a collection of foods that could be kept for a long time without the use of a fridge or freezer.

Discuss ways in which foods are prepared in the kitchen: using heat — baking, boiling, grilling, melting, frying; dissolving — sugar in drinks, making coffee, jelly, etc.; whipping and beating — cakes, meringues or milk shakes; chopping, grating, slicing.

⚠ Look at and handle different kitchen equipment. **Avoid anything sharp or potentially dangerous.** Look at and discuss the different materials — plastic, metal, glass, ceramic, wood. Read *In the Night Kitchen* by Maurice Sendak, Picture Puffins.

Vocabulary: Fresh, dried, frozen, preserved, tinned, sell-by date, refrigerator, fresh, stale, mouldy, dissolving, melting, boiling, baking, grilling, toasting, frying, scraping, chopping, whipping, beating, peeling, grating, shredding, plastic, metal, ceramic, wood.

Collections: Foodstuffs: flour, sugar, salt, baking powder, dried yeast, herbs, spices, dried pasta, beans, rice, fruit and seeds, custard powder, dried milk, dried vegetables, (peas, mushrooms, etc.); tinned food: fruit and vegetables; bottled food: jams, pickles, sauces, oil, vinegar.

Empty cartons from food which is refrigerated: margarine, yoghurt, milk, cream, cheese, prepared meals. Empty cartons from frozen foods: ice cream, vegetables, beefburgers.

Pictures of food or fresh food for temporary display.

Kitchen equipment: oven gloves, sieves, salad spinner, spoons, ladles, whisks, egg beaters, teapots, saucepans, rotary tin-openers, chopping boards, pot stands; kitchen 'papers': towel, foil, clingfilm, greaseproof, serviettes, doilies, filter paper; kitchen cleaning products: empty cartons of washing-up liquid, soap flakes, washing powder, washing liquid, fabric softener, sponges, dishcloths, pot scourers, brushes.

Kitchen surfaces: tiles, wood, laminates.

⚠ **Any food that is prepared and is intended to be eaten must be done so under hygienic conditions. Emphasise the need for cleanliness to avoid coming into contact with germs.**

Science Activities:

- Investigate dried foods. Look at dried peas, beans, mushrooms, pasta, etc. Why don't these ingredients go bad? What has happened to them? Cover with water and observe the changes. What happens to the shape, size, colour, texture? Does the smell change? How can we measure the changes? Draw around before/after. Weigh a certain amount before/after.
- What happens to food that is not kept refrigerated? Investigate what happens to milk that is left in different places. How long does it keep in the fridge compared with outside the fridge. What changes occur?
- Compare dried, frozen, tinned and fresh foods of the same kind — e.g. peas. Note differences in size, shape and colour.
- Which powders will dissolve in water? Find out whether baking powder, sugar, coffee, dried milk, flour, salt, cocoa, will dissolve in water. Try warm water. How many teaspoons of any one powder will dissolve in a cup of warm water? Does the level of the water change?

In the Kitchen

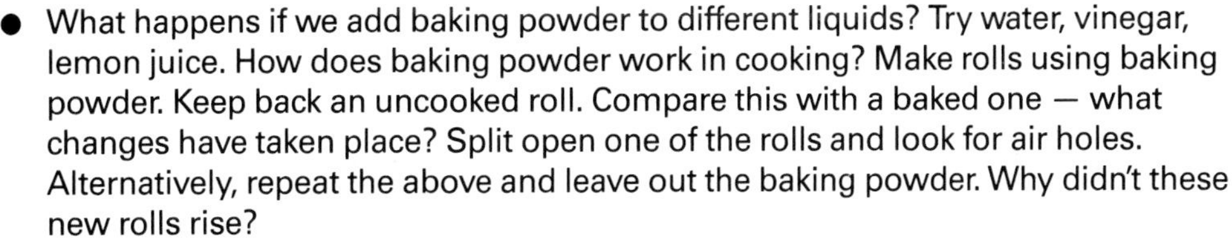

- What happens if we add baking powder to different liquids? Try water, vinegar, lemon juice. How does baking powder work in cooking? Make rolls using baking powder. Keep back an uncooked roll. Compare this with a baked one — what changes have taken place? Split open one of the rolls and look for air holes. Alternatively, repeat the above and leave out the baking powder. Why didn't these new rolls rise?
- **Why do we use yeast in bread-making?** Mix some yeast with a little warm water and sugar. Place in a bottle. Put an uninflated balloon over the neck of the bottle. Observe what happens. Look at slices of bread. Can you see the air bubbles?
- Make some toast. What happens to bread when it becomes toast? Place some newly toasted bread on a cold plate — what happens? Why does water appear on the plate? Weigh a piece of toast against a piece of bread from the same loaf. Which is heavier? Why? Why is toast hard and crunchy?
- Which type of milk makes the best butter? Obtain some different types of milk, e.g. skimmed, semi-skimmed and full cream. Observe carefully— how are the milks different/similar? Place equal amounts in identical jars. Shake jars continuously for the same amount of time, observing all the time. Which milk made the butter first? What was left after the butter was made? Can we turn butter back into milk? What does the butter feel like? What was taken from the skimmed and semi-skimmed milk?
- Make some chocolate crispy cakes. How can we use a block of hard chocolate in our recipe? How do we know that chocolate melts (sticky fingers)? When chocolate is melted, can we turn it back into hard chocolate again? Which other foods react in this way?

In the Kitchen

- Make jellies. Do small pieces of jelly melt more quickly than large pieces? What shapes can we make with jelly? How can we make jelly set more quickly? Try adding more or less water. What happens?
- How can we make our milk frothy? Try blowing through straws into individual milk shake drinks. How are the bubbles formed? Use an egg beater or whisk. Why does the milk go frothy? What other foods do we prepare by whisking or beating — (omelettes, sponge cakes, meringues)?
- Make ice lollies. How can we keep our ice lollies for as long as possible? What could we do if the freezer no longer worked? Investigate different insulating materials, e.g. newspaper, polystyrene chips, wood shavings, etc.
- Look at a food whisk. How does it work? Try turning the handle backwards and forwards — what happens?
- Which teapot keeps the tea warm for the longest time? Use ceramic, metal, and plastic teapots. Devise a way of testing which tea (or water) is warmest for the longest time.
- Why do we use washing-up liquid? Try washing greasy plates with/without washing-up liquid. Investigate what happens if you use warm or cold water. Why is warm water best for washing dishes?
- Investigate the properties of different kitchen papers — foil, baking parchment, greaseproof, clingfilm, kitchen towel, filter paper. Try dropping vegetable oil onto each one. What happens if water is dropped onto each?
- How can we mop up a spillage? Investigate different absorbent kitchen materials — sponges, tea towel, kitchen towel etc.
- Look at sponges. Can we make a sponge float/sink? What happens when you squeeze a sponge under water?
- How can we strain tea without a tea strainer? Investigate different materials that could be used — colanders, sieves, tea towel, sponges, coffee filter papers.
- Look at how a salad spinner works. How else could you dry lettuce without crushing it?
- Investigate kitchen surfaces — which ones are easy to clean and why?

Creative Activities:

- Bubble prints. Using a shallow container, mix ready mixed paint and water with washing-up liquid. Blow into it using a straw to make the bubbles rise up out of the container. Place paper over the bubbles to make a print. Try making multicoloured prints by over-printing using different colours.
- Vegetable oil and wax crayon 'stained glass' pictures. Draw a design onto paper. Go around outlines or details in black wax crayon. Colour in the design with coloured wax crayons. Dab vegetable oil evenly over the picture (use cotton wool). Allow to dry. Display designs on windows. Make circular or rectangular mounts out of black paper to give a frame to the picture.

- **Icing cakes and biscuits.** Prepare some coloured icing. Use either a piping bag, plastic syringes or spoons to decorate the cakes or biscuits. Make clown faces, smiling faces, stars, moons, etc. Add small sweets to give detail. Have a biscuit-icing competition.

In the Kitchen

- Starch resist. Mix flour and water to make a very smooth cream consistency. Place inside a clean washing-up liquid bottle. Tip the bottle upside down and make a design onto cloth (test on paper first). Allow to dry. When dry, crumple up the fabric and then smooth out. Paint on fabric dye, work into the cracks in the starch. When the fabric dye is dry, pick off the flour mix. A batik effect will emerge on the fabric (see photograph above). This also looks effective if a design is made on dark paper and simply left to dry.
- Make some natural dyes using red cabbage, onions, blackberries, etc. Dye some old white handkerchiefs. Try tie and dye.
- Use vegetables for printing. Try cutting in different ways to get a variety of blocks. It is very difficult to carve shapes out of vegetables, but circles, squares, elipses, semi-circles and triangles should be easy enough to obtain. **(No sharp knives).** Use ready mixed paint or water-based printing ink. Dab the block onto a piece of sponge or cloth before printing — this avoids over-loading the block with paint.
- Use teabags to give paper an aged look. Prepare some paper by dabbing over with a damp teabag. Draw a treasure map on the paper. Roll up and tie with a ribbon — use this as a prop for drama activities, etc.
- Make a collage using pasta or foods found in the kitchen store cupboard — breakfast cereals, nuts, seeds, etc.
- Design a machine that will do the washing-up, or make sandwiches, or pour drinks, etc.
- Observational drawings of kitchen equipment — especially beaters or other gadgets with working parts that are visible.

Shadows — Night and Day

Starting Points/ Discussion: Begin by sequencing a day's activities, starting from waking up and ending with going to bed. This could be in the form of an illustrated sequence. Discuss when the sun is shining and when it is not. The sun could be added to the illustrated sequence. Discuss sunrise and sunset, dusk and dawn. What happens to the sun at night? Illustrate this with the aid of a globe and light source (see Science Activities). What can we see in the sky at night-time? The moon shines by reflecting light. Discuss night workers: nurses, doctors, security guards, police, newspaper printers, etc. What animals are active at night (badgers, foxes, hedgehogs, bats, mice, etc.)?

How can we be seen in the dark? Discuss reflective and luminous colours for road safety. Why do we need sleep at night-time? Discuss the need for more rest for growing young creatures. Discuss times of going to bed and hours of sleep for babies/young children/adults.

Discuss the influence the sun has — it gives us light and warmth; it makes plants and (indirectly) animals grow; it dries up puddles; it is the source of energy for everything living on earth. Sunlight can pass through some materials (glass, clear plastic, etc.) but when it can't, shadows are formed. What are shadows like on a cloudy day?

Vocabulary: Nocturnal, sunrise, sunset, east, west, solar power, evaporation, reflective, luminous, shadow, transparent, translucent, opaque, midnight, twilight, dusk, midday, noon, axis.

Collections: Pictures of towns at night, night workers, nocturnal animals (have nocturnal animals to visit — hamsters); sunrises and sunsets; daytime pictures, e.g. holiday snaps; clothes and items with reflective and luminous strips; bicycle lamps, torches, candles, oil lamps, hollowed pumpkins, electric reading light; sunglasses, suntan lotion; a globe; pictures of warning lights, lighthouses; seedlings and pots; dark-loving minibeasts; sundial or picture of a sundial.

Science Activities:

- Why is it dark at night and light in the daytime? Obtain a globe that spins. Let the children experience spinning it. Can they find Britain? Can they find places they have visited? Find the North Pole and the South Pole. Using pieces of re-usable adhesive and a matchstick, indicate where they live. Shine a torch on to the globe so that the matchstick is in the light. Turn the globe slowly so that the matchstick is away from the light. Explain that the light represents the sun and that the earth spins round on its axis on its journey around the sun. As it spins one side faces the sun and the other faces into space and so on.
- How can we see in the dark? Investigate how torches work (see BATTERIES AND BULBS).
- What are shadows? On a sunny day go into the playground. Look at the shadows, make shadow towers, play shadow tag, make shadow statues. Are all the shadows facing the same way? Try going into a shady place — can shadows still be made? Why not?
- Do some minibeasts prefer the dark? Design a fair test to find out whether some creatures prefer the dark to light. A box with a cover over one end could be used together with a light source. Why do some creatures prefer dark places (damp, cool)?
- **Drying washing.** Why do we hang washing out to dry? Does it dry better in the sun or in the shade (see PUDDLES AND RAIN)?

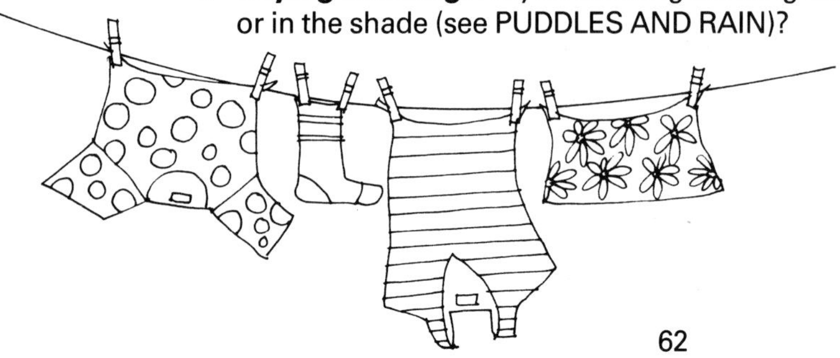

Shadows — Night and Day

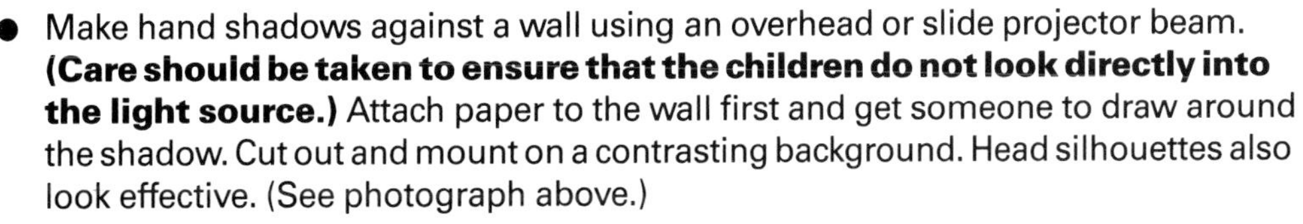

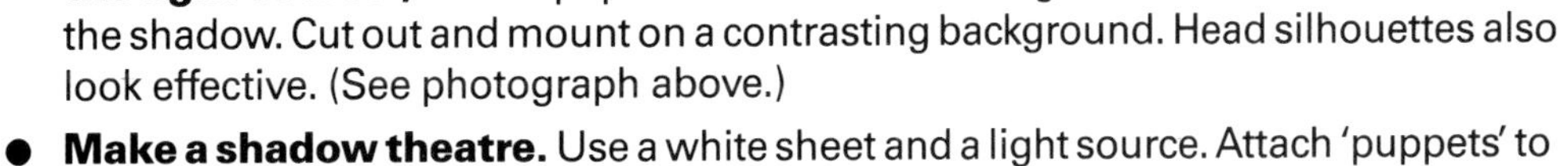

Creative Activities:

- Make hand shadows against a wall using an overhead or slide projector beam. **(Care should be taken to ensure that the children do not look directly into the light source.)** Attach paper to the wall first and get someone to draw around the shadow. Cut out and mount on a contrasting background. Head silhouettes also look effective. (See photograph above.)
- **Make a shadow theatre.** Use a white sheet and a light source. Attach 'puppets' to rulers, etc.
- Make shadow statues in the playground — sportsmen/women, monsters, spiky shapes, etc. Draw around on black paper — display with a caption.
- Sunset/sunrise pictures. This could be done using wax and wash technique. Tissue paper collage, collage using coloured paper from magazines, sponge painting, etc. Details could be added in the form of silhouettes either cut from black paper or drawn using wax crayons or oil pastels. Trees look particularly effective. Use travel brochures as a stimulus if the real thing is impossible to observe.
- Night-time lights. Use the wax and scratch technique — fairgrounds, bonfire night, car lights, traffic lights.
- Moonlight pictures. Paint a scene using only shades of grey and black and white. Use black paper for background, add details with white, grey and silver wax crayons.
- Moon pictures. Cut out a large circle of white paper — mount on to black or dark blue. Cut out symmetrical bats and stick on to the moon, or overlap into the dark blue.

Print stars using the end of a ruler with paint on black paper.

Add witch silhouettes, etc.

Waste

Starting Points/ Discussion: Carry out a litter survey of the school grounds. Discuss findings. Where were the litter bins situated? Were they full or empty? Discuss what happens to the waste we put in the rubbish bin. How often is it collected? Who collects it, and where do they take it? Why do we usually put rubbish in bags? What would happen if we didn't use bins? Discuss which types of rubbish can be burnt safely

⚠ **Talk about safety: some things we throw away can be dangerous** — for us and for wildlife — metal, glass, plastic bags. There is danger from rats, flies, germs, etc. on open rubbish dumps. Discuss liquid waste — drains, sewers, rivers, etc. Washing-up water is a waste material. What else is liquid waste? Encourage conservation awareness: discuss recycling and the need to avoid excessive waste.

Read *Dinosaurs and all that rubbish* by Michael Foreman, Picture Puffin.

Vocabulary: Rubbish, waste, litter, trash, junk, scrap, refuse collector, dustbin, decompose, rot, dissolve, rust, biodegradable, recycle, pollution, conservation, packaging, drains, sewers, rubbish dump, compost heap.

Collections: Household waste must be clean — no putrescent materials. Christmas and Easter are good times for collecting waste paper and packaging; used greetings cards, boxes, paper, ribbon, party cups and plates, sweet wrappers; refuse bags of different kinds, e.g. garden bags and bin liners; plastic food containers; plastic soft drink bottles; glass jars with lids; newspapers and magazines; recycled paper; biodegradable bags; tissues; kitchen roll; containers from a take-away meal; pictures of foxes raiding rubbish bins.

Science Activities:

- Collect waste materials from packed lunch boxes. Collect from several classes for a day. Sort into sets: plastics, card/paper, metal, etc. Weigh the packaging. Why do we need to wrap our food?
- What happens if rubbish is buried? Either dig holes in damp ground or fill trays/bowls/cut down plastic containers with soil. Place an item of rubbish into the soil and cover it up. Predict what will happen. Leave for a fortnight and compare with the same kinds of rubbish that have not been buried. Don't forget to label buried items.

⚠ **Plastic bags can be dangerous; they cause suffocation.**

- What happens to bread if left in a plastic bag? Obtain some samples of bread — wholemeal, white, multigrain, etc. Place into sturdy plastic bags and seal with a tie. Leave in a warm place — observe what happens. What is growing on the bread? Use hand lenses. Which bread goes mouldy the fastest?
- Try this with other foodstuffs, e.g. apple cores, cheese. Keep the bags tightly closed, or use glass jars with lids and seal with adhesive tape. Dispose of safely, after making observations.
- **How strong are refuse bags?** Obtain some refuse bags from different sources. Compare them for size, price, volume (e.g. how many shoe boxes will each one hold?). Devise a strength test. This could involve filling the bag with heavy objects and dragging it over a rough surface. Compare each bag for signs of wear. Does the plastic stretch, rip, tear?

Waste

- What happens to waste when put in water? Choose several items to test, e.g. plastic, metal, paper, cardboard, cloth, bread. Put items into small containers (rubbish collection will provide these). Cover with water. Observe carefully over several days. Try to find some items that will dissolve. Discuss the implications of this for rivers being used as rubbish tips.
- Why do we use plastic for packaging? Investigate the properties of different plastic containers. Will they tear, rip, bend? Are they heavy/light? Will they float/sink? Are they waterproof? What happens to plastic in hot water/in the freezer? Will it decay if we bury it?

Creative Activities:

- Make recycled paper.
- Using the paper collected by the class make papier-mâché models.
- Make transparent plastic sculptures. Collect as many different shaped plastic containers as possible. Arrange in shapes of animals, robots, factories, etc. Fill inside spaces with scrunched coloured Cellophane (see photograph above). The pieces can be joined with sticky tape or slits can be made, so that they slot together.

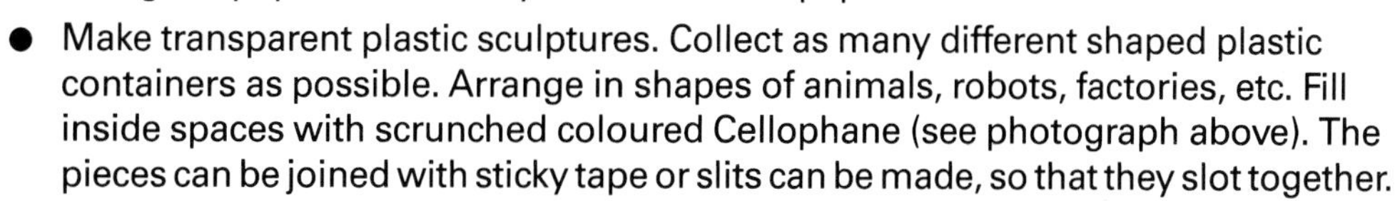

- Design a poster and a logo encouraging people to use recycling centres.
- Use waste materials for printing, e.g. rags, scrunched paper, cylinders, cuboids.
- Collect a large selection of boxes and containers — have a school sculpture that is changed every day.
- Make a collage using waste materials, e.g. ring pulls, milk bottle tops, silver foil, cartons, **paper cups (see line drawing).**

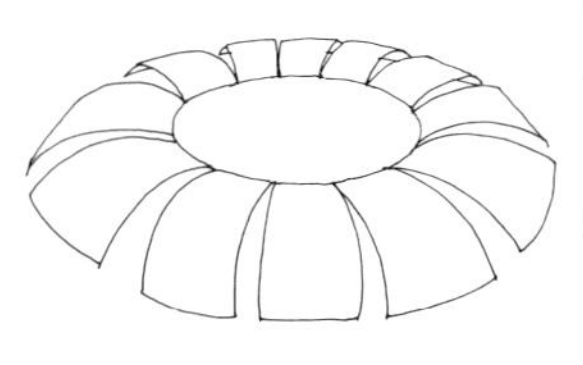

The Sky

Starting Points/ Discussion: What can we see, or might we see, in the sky in the daytime? Sun (moon), clouds, rainbows, lightning, areoplanes, helicopters, hot-air balloons, birds, smoke from chimneys, roof tops, weather vanes, tree tops.

⚠ **Never look directly at the sun.**

What can we see, or might we see, in the night sky? Stars, moon, clouds, satellites, lights of aeroplanes (UFOs!), comets and meteors. What colour can the sky be? Can we see where the sky begins? Go for a walk on to a high hill — or at the seaside. The sky comes down to 'touch' the ground. Have any of the children seen the clouds below them when flying in an aeroplane?

Discuss with the aid of a globe and light source (torch, lamp, etc.) why we can't see the sun in the night-time. The stars are always there — we can't see them because the sun's light is too bright in the daytime. Discuss the stars and constellation patterns. Discuss the turning of the earth and the earth's path around the sun. Introduce the moon, using a small sphere (tennis ball) and show that this travels around the earth. This will need supplementing with pictures of the earth/moon from space and relevant video material. Discuss satellite pictures for weather forecasts. Discuss our atmosphere and how this is like a protective bubble in which we can live — it gives us air to breathe and the weather, and it shields us from the sun's harmful rays. The weather aspect links with PUDDLES AND RAIN — the water cycle may be further reinforced and expanded upon. (Links also with SHADOWS — NIGHT AND DAY.)

Vocabulary: Pollution, cumulus, cirrus, stratus, moonlight, starlight, sunlight, constellation, planets, orbit, atmosphere, aeroplane, satellite, vapour trails, lightning, horizon.

Collections: Pictures of the earth and moon taken from space; pictures of the solar system showing the other planets and their orbits around the sun; pictures of constellations; pictures of meteors and comets; pictures and models of aeroplanes, hot-air balloons; models of birds — inflatable seagulls, etc.; pictures or reference books for identifying cloud types; globes and light sources; pictures of rainbows.

Science: Activities:

- Where does the sun rise and set? On a sunny day, as early as possible, go into the playground and stand with the sun directly behind. Draw around your own feet. Go back every hour to the same spot. Is the sun still directly behind if you stand in the same spot? Or note which window the sun shines through in the morning and watch what happens during the day. Explain this with the aid of a globe and light source.
- **Cloud watching.** Look out of the same window every day. Look at the clouds. Draw them and add colour. If possible identify them. Observe the weather conditions associated with different cloud types. Devise a means of recording the clouds for each day. If the clouds are moving — why is this happening?

	Monday	Tuesday	Wednesday	Thursday	Friday
a.m.					
p.m.					

Get your clouds here!
cumulus
stratus
cirrus
cumulo nimbus

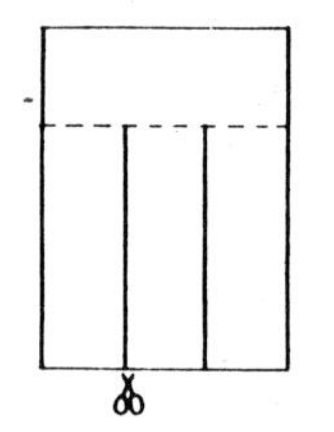

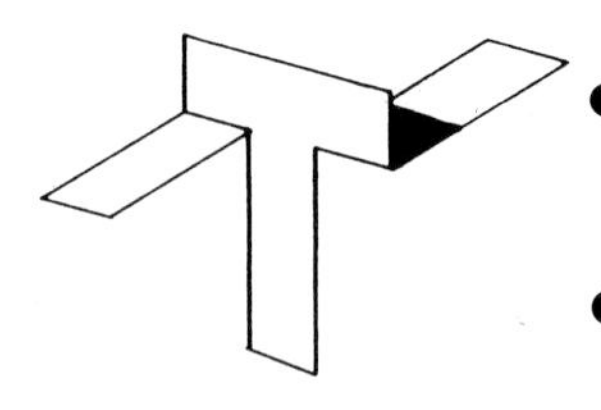

- **Make paper helicopters.** Investigate adding weights, e.g. paper clips to the end of your paper helicopter. Look at sycamore, maple and ash seeds. Toss them into the air and watch how they fall.
- Make paper gliders. Investigate different ways of folding — try putting weights on the nose. Look at pictures of real aircraft.

The Sky

- Investigate things falling from the sky. Throw balls into the air. Who can throw a ball ⚠ the highest? Throw other (safe) objects into the sky (bean bags, quoits, etc.). **Drop objects from different heights** (again, safe objects that will not harm children, or smash). Do all objects fall at the same rate? Do heavier objects fall more quickly? Investigate dropping paper in different ways — edge on, flat, folded, screwed up tightly. How can we slow things down that fall from the sky (parachutes)?
- Toy parachute-making. Which materials make the best parachute? Investigate size of canopy, length of strings, use different weights (toy cars, plasticine, cotton reels, etc.). Try dropping the parachute in a screwed up ball.

Creative Activities:

- Night sky. Splatter white paint on to heavy blue paper, add a silver moon cut from silver paper. Cut out shapes from black paper and glue on to the paper. Add touches of grey/light blue paint for moonlight.
- Looking up. Find a tree with spreading branches. Look up through the branches, observe the light of the sky and the dark tracery of leaf shapes. Draw the leaves with thick wax crayon and wash over with a sky wash. Or, using 'sky' colour paper, print the leaves and branches using corrugated card for branches, and sponges for the leaves.
- Foggy pictures. Use powder paint and cotton wool or dry sponges to make foggy pictures. Night-time pictures look particularly effective on black paper — lights of cars and street lamps drawn with circles of dry paint smudged outwards. Daytime pictures could show sunrises with trees and buildings in the mist.
- Hot-air balloons. Design a hot-air balloon — using collage materials. Sponge a sky background. Attach the balloon on bent card so that it stands out from the background.
- Cloud pictures. Devise ways of representing different cloud types — sponge print, fabric collage, splatter paint stencil (see photograph above).

Science from Stories

In the course of this text so far, stories have been suggested to supplement and develop particular topics. However, stories themselves can be starting points for science activities. Traditional stories and nursery rhymes together with contemporary picture books contain a wealth of ideas for scientific investigations, setting challenges and providing opportunities for problem-solving. The following suggestions are the result of discussions and observations made by five and six year olds during class storytimes.

ROSIE'S WALK BY PAT HUTCHINS, PICTURE PUFFIN

- Why does the rake hit the fox? What makes it spring up into the air? Investigate using models or toy rakes. Talk about safety — do not leave garden tools on the ground!
- Why does the haycock (haystack) collapse when the fox jumps on to it, but not when Rosie walks over it?
- Why does the flour sack come crashing down? Investigate pulley systems and hoists. Make your own flour bomb!
- What starts the cart moving? Why does it roll? Make a wheeled vehicle with construction toys and investigate rolling, gradient, inertia, etc.
- The beehives fall over when the cart hits them. Can you build a structure of dominoes or wooden building bricks that fall over like this, so that when one falls it topples the next, which knocks over the next, and so on?
- Make a working model of Rosie's farmyard to illustrate her walk.
- **Make a moving model of Rosie.**

'WOULD YOU RATHER . . .' BY JOHN BURNINGHAM, JONATHAN CAPE

This is a storybook packed full of starting points for discussion and classwork.

- What problems would you encounter if your house was surrounded by water, snow or jungle? How could you escape from your house? Talk about boat or raft-making. Which household objects or pieces of furniture could you use? Consider both shape and materials. How could you clear or melt the snow? How could you travel over the snow, make snow shoes, skis or a sledge using items available in your house? Talk about problems of travelling through the jungle. Could you swing from tree to tree on jungle vines? Could you make a rope bridge and rope ladders?
- Design clothes suitable for being covered in jam, soaked with water, or being pulled through the mud by a dog. Would the clothes have to be waterproof, washable, tough, padded, etc.?
- What problems would you encounter having 'breakfast in a balloon, supper in a castle or tea on the river?' How could you stop the ducks from eating your tea? How could you keep the food hot, or fresh? Devise suitable menus for your meals.
- What is a fog? How could you find your way home in a fog? How could you avoid going round in circles in a forest? Where might you find water in a desert? How could you attract attention out at sea, on a raft?

Science from Stories

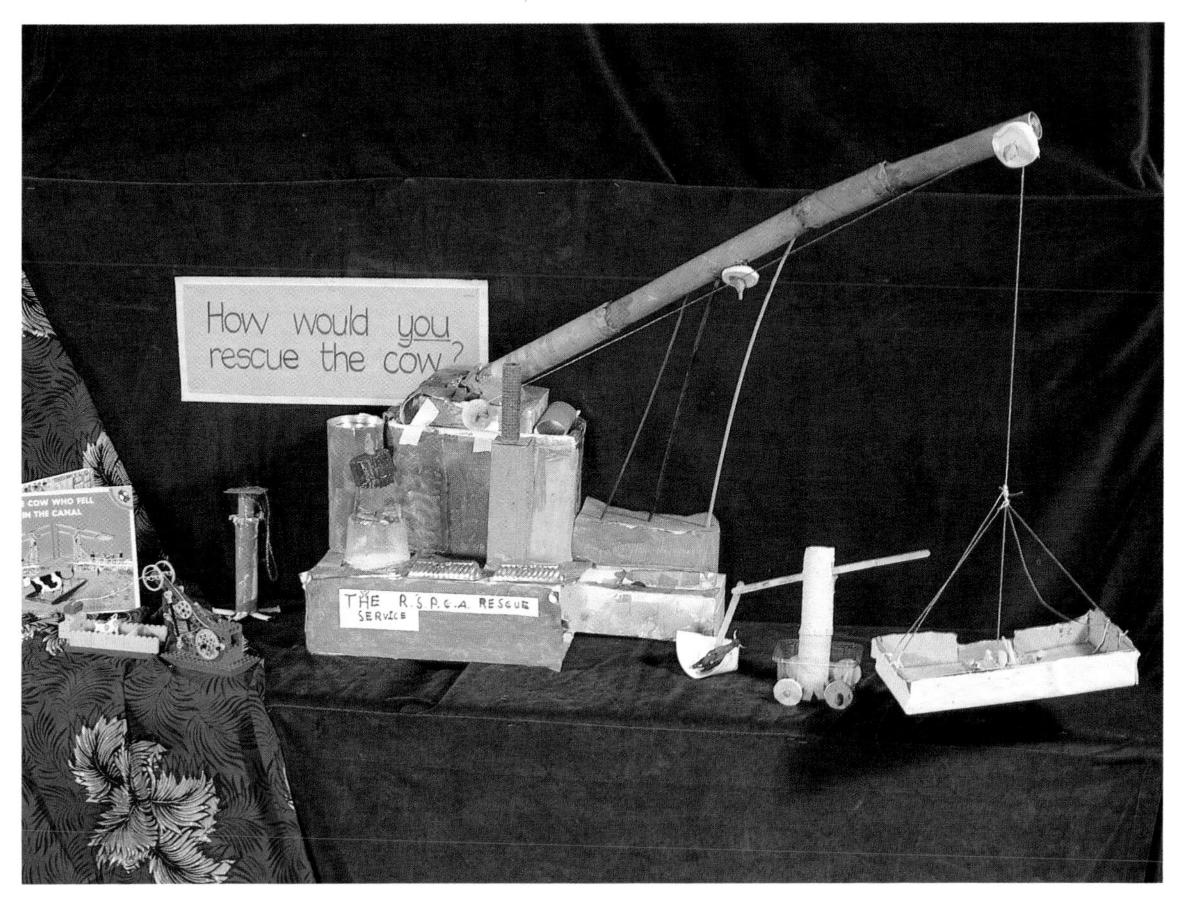

ON FRIDAY SOMETHING FUNNY HAPPENED BY JOHN PRATER, PICTURE PUFFIN

Each day of mischief provides an opportunity for discussion.

- At the supermarket: look at the way in which the trolley moves. What is different about supermarket trolley wheels? Look at the way in which the food is stacked and balanced.
- At the park: investigate the movement of swings. Find out about slides, roundabouts, etc.
- Doing the washing: experiment with bubbles, soaps and drying things.
- Playing with friends: investigate simple musical instruments. Talk about tricycles, bicycles, stability and safety on the roads.

THE COW WHO FELL IN THE CANAL BY PHYLLIS KRASILOVSKY, PICTURE PUFFIN

The remarkable illustrations by Peter Spier contain a wealth of scientific detail.

- Look for different kinds of bridges.
- Look for different ways of lifting things — there are pulleys and hoists used for lifting cheeses, and raising flags. There is a bridge that lifts — how do you think it is raised?
- Set a challenge. How could Hendrika be lifted out of the canal safely? Make moving, lifting machines (see photograph above).
- Investigate how wind power can be harnessed.

Science from Stories

THE SNOWMAN BY RAYMOND BRIGGS, PICTURE PUFFIN

- How would you dress for playing in the snow? Investigate how some clothes keep you warm and dry.
- How many sources of heat did the snowman find inside the house? How many cold things did he find?
- Investigate — the squirting power of a squeezy bottle
  — the rolling of a skateboard
  — the spring of a punch ball
  — the beam of a torch
  — playing with balloons.
- What kinds of food do you think a snowman would like? Plan a menu for a snowman's teaparty.
- How could you preserve a snowman and stop him from melting?

THE LIGHTHOUSE KEEPER'S LUNCH BY RONDA AND DAVID ARMITAGE, PICTURE PUFFIN

- How can you make a basket travel across the classroom on a pulley?
- Make a model lighthouse using a simple electric circuit to light the lamp. Use a switching device to make the light flash.
- Can you devise a plan to baffle the seagulls? e.g. make a 'seagull-proof' basket or container for the lighthouse keeper's lunch.

The Lighthouse Keeper's Catastrophe by the same authors provides a good follow-on for these activities.

- Why could Mr Grinling have a holiday when the sun was shining? Discuss weather forecasting. Draw a very simple weather map to show that wind and rain are on the way.
- Why did Mrs Grinling weigh the sack and send it down the wire first? Find out about a 'breeches-buoy'.

Resource Books

LOOK! Primary Science Project, *A first LOOK!, LOOK! Teachers Guide A,* by Cyril Gilbert and Peter Matthews, Oliver and Boyd.

Science Through Infants' Topics, Teachers books, by Enid Hargrave and Janet Brooks, Longman.

Painting with Energy, by Henry Pluckrose, The Electricity Council.

Science 5/13, Schools Council, Macdonald.

Learning Through Science, Macdonald Education.

An Early Start to Science, by Roy Richards, Margaret Collis and Doug Kincaid, Macdonald Education.

Bright Ideas for Science, Scholastic Publications, Wardlock Educational/Scholastic.

US, Sussex Primary Science Study Group, Osmiroid Educational.

Bridges (Science Explorers), by Annabelle Dixon, Adam and Charles Black.

Thinking Science, by Linda Allison and David Katz, Off Beat Books, Cambridge University Press.

Beastly Neighbours, by Mollie Rights, Off Beat Books, Cambridge University Press.

The Amazing Science Amusement Arcade, Ontario Science Centre, Cambridge University Press.

Linnea's Windowsill Garden, by Christina Björk and Lena Anderson, Raben and Sjögren Books.

Fun with Science (series), by Brenda Walpole, Kingfisher Books (Grisewood and Dempsey Limited).

The Knowhow Books (series), Usborne Publishing Limited.

Usborne First Nature Books, Usborne.

Usborne First Science and Maths, Usborne.

Starters Science Books, Basic Starters, Macdonald.

Thinkabout, Teachers' Notes, BBC Books.

Althea's Nature Series, Dinosaur Publications.

The Sainsbury Book of Children's Party Cooking, by Carol Handslip, Cathay Books.

Health for Life 1, The Health Education Authority's Primary School Project, Nelson.

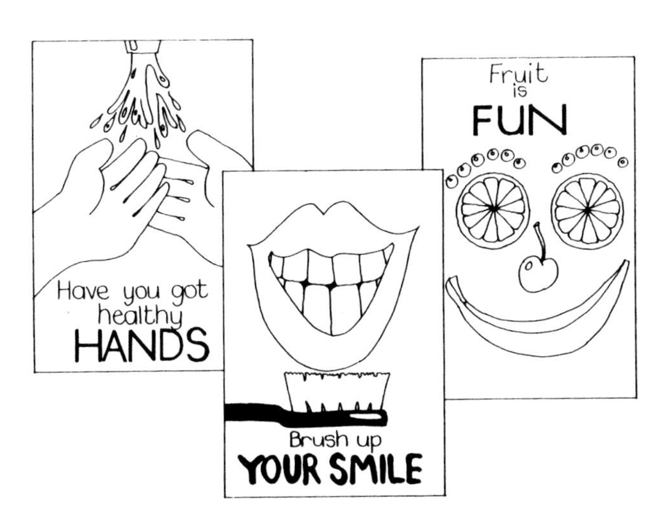

For details of further Belair Publications
please write to:

BELAIR PUBLICATIONS LTD.,
P.O. Box 12,
TWICKENHAM
TW1 2QL,
England.